WITHDRAWN
UTSA LIBRARIES

CLAYS AND CERAMIC RAW MATERIALS

SECOND EDITION

CLAYS AND CERAMIC RAW MATERIALS

SECOND EDITION

W. E. WORRALL
M.Sc., Ph.D., M.R.S.C., C.Chem.

*Department of Ceramics,
Houldsworth School of Applied Science,
University of Leeds, Leeds, UK*

ELSEVIER APPLIED SCIENCE PUBLISHERS
LONDON and NEW YORK

ELSEVIER APPLIED SCIENCE PUBLISHERS LTD
Crown House, Linton Road, Barking, Essex IG11 8JU, England

Sole Distributor in the USA and Canada
ELSEVIER SCIENCE PUBLISHING CO., INC.
52 Vanderbilt Avenue, New York, NY 10017, USA

First edition 1975
Second edition 1986

WITH 70 ILLUSTRATIONS AND 37 TABLES

© ELSEVIER APPLIED SCIENCE PUBLISHERS LTD 1986

British Library Cataloguing in Publication Data

Worrall, W. E.
Clays and ceramic raw materials.—2nd ed.
1. Clay—Analysis   2. Ceramic materials—Analysis
I. Title
666   TP811

Library of Congress Cataloging-in-Publication Data

Worrall, W. E.
Clays and ceramic raw materials.

Includes bibliographies and index.
1. Clay. 2. Ceramic materials. I. Title.
TP811.W65   1986   666   86-11661

ISBN 1-85166-004-6

Special regulations for readers in the USA
This publication has been registered with the Copyright Clearance Center Inc. (CCC), Salem, Massachusetts. Information can be obtained from the CCC about conditions under which photocopies of parts of this publication may be made in the USA. All other copyright questions, including photocopying outside of the USA, should be referred to the publisher.

All rights reserved. No part of this publication may be reproduced, stored in a retrieval system, or transmitted in any form or by any means, electronic, mechanical, photocopying, recording, or otherwise, without prior written permission of the publisher.

Phototypesetting by Keyset Composition, Colchester, Essex
Printed in Great Britain by Galliard (Printers) Ltd, Great Yarmouth

Preface to the Second Edition

Over the past ten years, remarkable progress has been made in our understanding of clays. Improvements in instrumentation have facilitated detailed research into the structures and interrelationships of the kaolin group of minerals by X-rays, infra-red spectroscopy, electron microscopy and thermal methods. In particular, both X-ray diffraction and infra-red spectroscopy have been used for identifying and quantifying disorder in kaolinites. The relatively new method of Mössbauer spectroscopy has furnished direct evidence for the substitution of aluminium by iron in kaolinites whilst the subject of intercalation of clay minerals has also received much attention. In recent years much greater emphasis has been placed on the preparation and purification of clay specimens for physical tests; in addition to the well-established physical separation methods based on size fractionation, heavy liquid separation and magnetic separation, there are a steadily growing number of chemical methods.

Even more recently, the colloidal properties of non-clay materials have received considerable attention, due to renewed interest in wet methods of fabrication; it is anticipated that the sections on colloids and rheology may provide a useful introduction to those entering this new field.

In this second edition, the opportunity has been taken to incorporate much of this new information. The sections on crystallinity and structure of kaolinites and illites, the stability of colloids, the rheology of montmorillonites, intercalation and drying have all been expanded, whilst a new section on chemical methods of separation has been added. In response to suggestions made, the bibliography has now been considerably expanded; in addition to an extended reading list, specific references to appropriate papers are now included at the end of most chapters. As in the first edition, the subject matter has been confined to 'raw' materials in the sense of those extracted from the earth, despite the rapid development in recent years of synthetic 'raw' materials for special ceramics. It is proposed to cover these important synthetic materials in a future separate publication.

Preface to the First Edition

Despite the enormous and growing importance of ceramic materials in many industries, there are few books on clays and ceramic raw materials of a suitable standard for scientists and technologists wishing to gain a rapid insight into this subject. The present work attempts to bridge the gap between the very elementary treatment to be found in student textbooks and the specialised monographs and scientific papers.

In writing the book the author has drawn on the very extensive literature available and to a large extent on his own research and teaching experience, particularly on the structure and properties of clays. The aim here has been to present a logical and coherent account of clays and other raw materials, without burdening the reader with a mass of unconnected data.

It has been assumed that the reader has a working knowledge of chemistry and physics; the introductory chapters on structure serve to remind him of the basic facts on which the subsequent chapters are based and are in no way intended as a substitute for the existing texts on physical and inorganic chemistry. For those requiring more detailed information on the subject matter, a reading list has been included at the end of each chapter.

As far as is practicable, SI units have been used in the text, though there are inevitably certain exceptions, such as figures from papers published before the introduction of the new system. Another difficulty arises in the units of cation exchange capacity; the latter is still commonly expressed in millequivalents per 100 grams (meq 100 g^{-1}), this being a very convenient unit. The practice has been adhered to in the present text, in the absence of a satisfactory alternative.

The author gratefully acknowledges the generous assistance given in preparing the text by the British Ceramic Society, the Institute of Ceramics, the British Ceramic Research Association and by his colleagues in the Department of Ceramics, University of Leeds.

Contents

Preface to the Second Edition v

Preface to the First Edition vi

Chapter 1 *Fundamental Principles of Structure* . . . 1
   Introduction: the Range and Scope of
   Ceramic Raw Materials 1
   Lattices 2
   Crystals 4
   Bonding in Crystals 6

Chapter 2 *Silica* 14
   Quartz 15
   Structures of Cristobalite and Tridymite 16
   Conversions 17
   Inversions of the Forms of Silica 18
   Other Forms of Silica 20
   Physical Properties of the Various Forms of Silica 22
   Chemical Properties 23
   Occurrence of Silica 24

Chapter 3 *Structure of the Main Types of Clay Minerals* . 27
   General Aspects of Silicate Structures 27
   The Clay Minerals 30
   Framework Structures 45
   Mixed-Layer Structures 46

Chapter 4 *Geology of the Clays* 48
   Composition of Igneous Rocks 49
   Composition of Sedimentary Rocks 49
   Occurrence and Classification of Clays 55

Chapter 5   Properties of Clay–Water Systems ... 89
   The Importance of Clay–Water Systems ... 89
   Colloids ... 89
   Cation Exchange Reactions ... 103
   Anion Exchange ... 109
   Deflocculation and Flocculation of Clays ... 110
   Physical Adsorption by Clays ... 115
   Intercalation Compounds ... 122
   Hydrogen Clays and pH Titration Curves ... 123

Chapter 6   The Rheology of Clay–Water Systems ... 127
   The Importance of Rheological Studies ... 127
   Definition of Viscosity ... 127
   Measurement of Viscosity ... 129
   Non-Newtonian Flow ... 132

Chapter 7   The Plasticity of Clays ... 147
   Definition of Plasticity ... 147
   Plastic Flow ... 147
   Stress–Strain Measurements ... 149
   The Atterberg Plasticity Index ... 154
   Theory of Plasticity ... 157
   Application of the Rheological and Other Properties of Clay ... 160

Chapter 8   The Effect of Heat on Clays ... 167
   The Drying of Clays ... 167
   Thermal Decomposition of Clays ... 172

Chapter 9   Methods Used for the Identification and
              Characterisation of Clays ... 177
   Chemical Analysis ... 177
   Chemical Methods of Separation ... 185
   X-Ray Diffraction ... 188
   Differential Thermal Analysis ... 192
   Thermogravimetric Analysis ... 196
   Infra-red Absorption Spectroscopy ... 197
   Optical Microscopy ... 199
   Electron Microscopy ... 200
   Mössbauer Spectroscopy ... 201

Chapter 10   Refractory Raw Materials ... 204
   Alumina ... 204
   Refractory Silicates ... 208
   Steatite ... 209
   Magnesite ... 211
   Dolomite ... 214

Chrome	215
Rutile	218
Zirconia and Zircon	219

Chapter 11 *Miscellaneous Raw Materials* 220
    Fluxes 220
    Soda- and Potash-bearing Minerals 220
    Bone 225
    Plaster of Paris 227

Index 233

Chapter 1

Fundamental Principles of Structure

INTRODUCTION: THE RANGE AND SCOPE OF CERAMIC RAW MATERIALS

The oldest ceramic raw material is undoubtedly clay. Clay has been defined as an earth that forms a coherent, sticky mass when mixed with water; when wet, this mass is readily mouldable, but if dried it becomes hard and brittle and retains its shape. Moreover, if heated to redness, it becomes still harder and is no longer susceptible to the action of water. Such a material clearly lends itself to the making of articles of all shapes.

Clay may take various forms; it is easily recognised as the sticky, tenacious constituent of soil, but it frequently occurs as a rock which, owing to compression, is so hard and compacted that it is not initially plastic and is almost impermeable to water. Such rocks can, however, be rendered plastic by suitable treatment. Like all rocks, these clay formations contain a number of different minerals, but those responsible for the clay-like properties are relatively few in number and are easily identified; the remainder may be regarded as impurities. In general, the clay minerals, as they are called, fall into either of two main groups, the kaolinites and the montmorillonites, the former being the more important industrially.

Although the earliest ceramic articles were probably made entirely from natural clay, from very early times additions to it of other minerals are known to have been made. At the present time, the chief raw materials used in conjunction with clay for making pottery are silica and certain alkali-bearing materials used as fluxes; these materials clearly merit attention. In the refractories industry, increasing demand for specialised refractories has resulted in products containing little or no clay, such as alumina, magnesite and chrome refractories, which are also classed as 'ceramics' because generally they are shaped whilst moist and then fired in order to harden them. All these refractory products are made more or less directly from raw materials, the occurrence and properties of which are therefore of considerable interest.

In recent years a wide variety of inorganic, non-metallic materials has been developed for the electrical, nuclear power and engineering industries. These products are commonly fabricated as powders and afterwards heat-treated to consolidate them; hence, they too can be regarded as ceramic materials. Examples of the raw materials used in manufacturing these products are: rutile, a form of titanium dioxide used for ferro-electric materials; steatite or talc, for electrical insulators; alumina, zirconia, thoria and beryllia for refractories and electrical insulators; uranium oxide as a nuclear fuel element; and various carbides and nitrides as abrasives, insulators, heating elements and occasionally as engineering construction materials.

LATTICES

The development of X-ray diffraction and other techniques in comparatively recent times has revealed that the majority of solids are crystalline, the principal exceptions being glasses and gels. The solid crystals are composed of a regular array of atoms called a *lattice*. Each lattice contains a definite 'pattern' which is repeated throughout in a systematic manner. Figure 1 is a two-dimensional representation of a lattice containing two

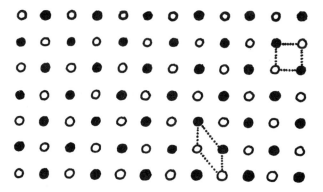

Fig. 1. A simple lattice.

kinds of atom only. The smallest group of atoms capable of representing the 'repeat pattern' is known as a *unit cell*. In Fig. 1 two possible unit cells are indicated in outline; from this it will be clear that there may be several alternative unit cells for any given lattice, the simplest usually being chosen for convenience. It should also be clear that in lattice compounds it is incorrect to speak of 'molecules'; the smallest entity in such compounds is the unit cell.

For a three-dimensional lattice, the unit cells are simple geometrical

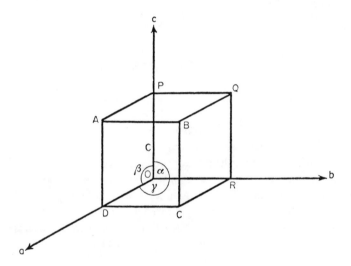

Fig. 2. Crystallographic axes.

solid figures, which are characterised by the lengths of three sides, a, b and c, and three angles, α, β and γ (Fig. 2). It can be shown geometrically that there are fourteen possible types of unit cell, sometimes referred to as the Bravais space lattices, which are subdivisions of the seven main crystal systems (Table 1). The most symmetrical of these is the cube, for which $a = b = c$, and $\alpha = \beta = \gamma = 90°$; and the least symmetrical is the triclinic, for which the three sides and the three angles are all unequal. Although the unit cell is the simplest representative unit, crystals can have more complicated shapes than their respective unit cells, because their faces may consist of any plane of atoms and need not coincide with the unit cell faces.

Table 1
The Principal Types of Unit Cell

Unit cell type	Unit cell dimensions	
Cubic	$a = b = c$;	$\alpha = \beta = \gamma = 90°$
Tetragonal	$a = b$; c;	$\alpha = \beta = \gamma = 90°$
Orthorhombic	a, b, c;	$\alpha = \beta = \gamma = 90°$
Monoclinic	a, b, c;	β; $\alpha = \gamma = 90°$
Triclinic	a, b, c;	α, β, γ
Hexagonal	$a = b$; c;	$\gamma = 120°$; $\alpha = \beta = 90°$
Rhombohedral	$a = b = c$;	α; $\beta = \gamma$

CRYSTALS

Even before the discovery of X-rays, extensive studies by visual methods had enabled crystallographers to classify crystals according to their *symmetry*, into seven distinct *systems*. A solid object is considered to be symmetrical if it can be split into two identical portions; the more often this process can be repeated, the greater the symmetry. A more precise way of assessing symmetry is to choose any convenient axis about which the crystal may be rotated. A highly symmetrical crystal will present the same appearance several times on being rotated through 360°. In Fig. 3, a cube is imagined to be rotated about an axis XY. It will be seen that as it is rotated through a complete revolution, four identical faces present themselves, that is, XY is an axis of *fourfold symmetry*. It will also be clear that there are three such fourfold axes of symmetry.

Rotation, however, is only one of several *symmetry operations*, as they are called. Another example is a *mirror plane* of symmetry, which divides the crystal into two halves, one of which is a mirror image of the other. Again, a *centre of symmetry* is a point within the crystal, such that any line drawn through it intersects the crystal surface at equal distances on either side. Taking into account all the elements of symmetry, the seven main crystal systems can be further subdivided, making in all thirty-two possible classes, or *point groups*, as they are termed.

These symmetry elements, based on external crystal form, are not sufficient, however, to define all crystals completely; it is necessary in addition to take their *internal symmetry* into account. Because lattices are in general three-dimensional, and can consist of many different kinds of atom, there are fourteen possible types of unit cell, as we have noted.

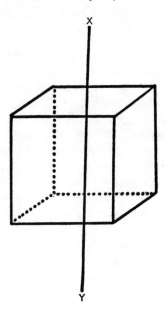

Fig. 3. Crystal symmetry.

Table 2
Crystal Systems and Space Groups
(From Glasstone's *Textbook of Physical Chemistry*,
copyright 1946, D. Van Nostrand Co. Inc., Princeton, N.J.)

System	Number of space groups
Cubic	36
Tetragonal	68
Orthorhombic	59
Monoclinic	13
Triclinic	2
Hexagonal	22
Trigonal or rhombohedral	30

Taking all the additional elements of symmetry into account, as well as the possibility of unlike atoms, the total number of symmetry operations or *space groups* is increased to 230. A list of the seven crystal systems, with their respective subdivisions into space groups, is given in Table 2.

Crystallographic axes

It is necessary to refer the various planes in a crystal or unit cell to three crystallographic axes, usually denoted by a, b and c. Where possible, these axes are, for convenience, selected to coincide with the edges of the crystal faces or with the axes of symmetry, and are at right angles to each other. The selection of crystallographic axes is best understood by considering a cubic lattice or cubic crystal ABCDOPQR (Fig. 2). Taking the point O as origin, three axes, Oa, Ob and Oc, are selected to coincide with the edges of the cube. Although these axes are mutually at right angles, with less symmetrical crystals this need not be so.

The various planes or faces of the crystal can now be referred to the three axes. Consider first the plane PDR, lying diagonally across the cube. This plane makes intercepts OD, OR and OP on the three axes respectively. The actual lengths of these intercepts are not important, but only the ratio between them. Now consider the intercept lengths OD, OR and OP each as *unit lengths* of their *respective axes*. The intercept ratios on the respective three axes are then 1:1:1. The Law of Rational Indices states that the intercepts made by any other crystal plane on the three axes will bear a simple ratio to OD, OR and OP respectively.

The reciprocals of the three intercepts made by any crystal plane, taking OD, OR and OP as units, are known as the *Miller Indices*. Thus, the Miller Indices of the plane PDR are respectively 1/1, 1/1, 1/1, and this plane is therefore said to be a 1:1:1 plane. The reason for using reciprocals becomes clear when the crystal plane ABCD is considered. This plane makes the intercept OD or unity on the a-axis, but since it is parallel to the b- and c-axes, its intercepts on these are infinite. Therefore the intercepts are $1:\infty:\infty$, and the Miller Indices are the respective reciprocals of these, namely 1:0:0, and ABCD is spoken of as a 1:0:0 plane. Similarly, APQB is a 0:0:1 plane. Planes having higher indices can be found in more complicated unit cells.

BONDING IN CRYSTALS

The electrovalent bond

Atoms with completed electron shells, as in the inert gases helium, neon and argon, are very stable and do not react with other elements to any great extent. However, atoms with incomplete electron shells tend either to gain or lose electrons to acquire a more stable configuration; this they can sometimes do by acquiring extra electrons from other atoms or by donating them to other atoms.

Consider, for example, the electronic structure of the fluorine atom. The first shell ($n = 1$) is completely filled; the second shell, and in particular the subgroup 2p, is incomplete, because quantum theory allows a total of six 2p atoms, whereas only five are present in fluorine. Therefore, the fluorine atom has a strong tendency to acquire an extra electron to complete the shell. It cannot readily accommodate more than one extra electron, because any in excess of this would have to be accommodated in the next shell ($n = 3$) and this is too far from the nucleus to be stable. Of course, when the fluorine atom does acquire its extra electron, there is an excess of negative charge and therefore a negative ion (anion) F^- is formed. This extra electron must, obviously, come from another atom that readily parts with an extra electron. Such an atom is lithium, which has only one electron in the second ($n = 2$) shell and so readily donates this odd electron to a fluorine atom, resulting in a complete ($n = 1$) shell and no electrons in the outer shells (i.e. a much more stable arrangement). In this case, the loss of one electron means that the lithium atom now has an overall positive charge and has become a positive ion (cation) Li^+. The most important result of this interchange, however, is that the two ions, Li^+ and F^-, are now strongly bound together by electrostatic attraction and the compound lithium fluoride, LiF, is formed. The bond holding the two ions together is called an *electrovalent bond*. Ions held together in this manner have a great tendency to form *crystal lattices*; it is seldom that individual molecules are formed, except in the gaseous state.

In general, all the alkali metals (Li, Na, K, Rb, Cs) form electrovalent compounds like LiF, with any one of the halogens (F, Cl, Br, I). All are typical *salts*, and are also *electrolytes*, because when dissolved in water they split up into their respective ions, which are then capable of migrating when a potential difference is applied to the solution.

Electronegativity

Atoms that part readily with electrons to form cations are said to be *electropositive*, whilst those that readily attract additional electrons are said to be *electronegative*. Fluorine is the most electronegative element, caesium the most electropositive; but between these two extremes are varying degrees of electropositive or electronegative character, governed by the relative stabilities of the incomplete electron shells. A scale of *electronegativity* has been devised to include all the elements; thus, on this scale the halogens have a high value, whilst the alkali metals have a very low value. Elements such as boron, carbon and nitrogen have intermediate values. Since the formation of an electrovalent bond involves the transfer of an electron from one atom to another, it is clear that such bonds will be formed most readily between elements of widely differing electronegativities. For elements having similar electronegativities, an entirely different kind of bond is formed, as described below.

The covalent bond

It is well known that bonds can be formed between atoms of the same element, as in the hydrogen molecule, H_2, and other gases. Clearly, there can be no question of direct electron transfer in such a molecule, hence the bond cannot be electrovalent. It is termed a *covalent bond*, formed by the sharing of electrons of two or more atoms. For example, the hydrogen atom can accommodate one more electron in the first shell, since two 1s electrons are permissible; this can be accomplished if one electron is 'shared' with another hydrogen atom. It can be shown by wave-mechanical theory that a strong bond is formed between two hydrogen atoms if (a) the two electrons are 'shared' by both atoms and (b) if the electron spins are in opposite senses (i.e. the spin quantum numbers must be of opposite sign). Since, in a mass of hydrogen gas, there is an equal chance of a given atom having positive or negative spin, there will in general be an equal number of both kinds of atom and stable H_2 molecules are therefore formed. This is an example of the *covalent bond*.

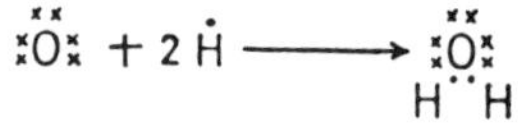

$$:\overset{\times\times}{\underset{\times\times}{O}}: + 2\,\dot{H} \longrightarrow \overset{\times\times}{\underset{H\,\,\,H}{:\overset{..}{O}:}}$$

Fig. 4. Covalent bonding.

Covalent bonds between other (unlike) atoms are formed in general by the sharing of pairs of electrons with opposite spins, e.g. the covalent bond in H_2O. From observations of atomic spectra, it is known that the oxygen atom has only *four* 2p electrons. Since the maximum permissible number of 2p electrons is six, the oxygen atom can acquire a more stable state by sharing a pair of electrons with each of two hydrogen atoms (Fig. 4). For clarity, the electrons originally associated with the hydrogen atoms are shown as dots; those associated with the oxygen atoms are shown as small crosses. The two 'shared pairs' of electrons are indicated by heavier type. By this mutual sharing, both the oxygen and hydrogen atoms have become more stable because the shared pairs of electrons have opposing spins. Oxygen cannot have a valency greater than two, because the second shell ($n = 2$) is now completed. On the other hand, it cannot have a valency less than two because there are always two 2p electrons with 'unpaired' spins. Many more compounds having covalent bonds can be formulated in the same way, e.g. LiH, NH_3, CH_4, CCl_4 and HF.

The polarity of bonds

When two atoms of the same element combine to form a molecule, as in H_2, the shared electrons are evenly distributed or symmetrical, there being no resultant charge on either atom. When two atoms of different electronegativity combine, however, the electrons are not quite evenly shared but tend to be closer to one atom than the other. Thus, in HCl, the hydrogen atom, being more electropositive than chlorine, tends to repel the shared

Fig. 5. Polarity of bonds.

electron and acquires a resultant small positive charge; similarly, the chlorine atom acquires a small negative charge (Fig. 5). Then if the fractional charges on the atoms are $+q$ and $-q$ respectively and r is the distance between them, the HCl molecule has a resultant electrical moment, called a *dipole moment*, measured by $q \times r$. A molecule of this type is known as a *dipole* or *polar molecule*. The dipole moment of a molecule depends largely on the value of q, which in turn depends on the difference in electronegativity. As the latter increases, the charge q also increases, and in an extreme case the bond between the atoms can no longer be regarded as covalent but becomes wholly electrovalent, with q equal in magnitude to the charge on the respective ions. That is to say, there is no sharp distinction between covalent and electrovalent bonds, all intermediate types being possible. Thus, depending on the difference in electronegativity, the bond between two unlike atoms may be regarded as possessing a proportion of both ionic and covalent character. This is illustrated in Table 3. From this it will be evident that the silicon–oxygen

Table 3
Percentage Ionic Character of Bonds

Bond	Ionic character (%)	Covalent character (%)
Na—F	90	10
Al—O	63	37
Si—O	50	50
Si—F	70	30
Si—Cl	30	70
Mg—O	73	27
B—O	44	56

bond, an important one from the ceramic point of view, has a considerable degree of ionic character, although it is often thought of as covalent. This ionic character has a considerable bearing on the current view of silicate structures, as referred to later.

The hydrogen bond

Although hydrogen is normally regarded as monovalent, it can nevertheless act as a 'bridging atom' between certain other atoms, notably oxygen and fluorine, as shown in Fig. 6. The hydrogen bond is particularly

Fig. 6. The hydrogen bond.

important in hydroxylic liquids like water, because it causes them to associate, forming complex molecules of type $(H_2O)_3$. It is best imagined by supposing that a bonding electron from hydrogen jumps rapidly from one oxygen to another, thus bonding itself to each oxygen in turn. This is depicted below as two forms, one changing rapidly into another and back again. The Wave Mechanical Theory enables us to formulate two or more simultaneous modes of vibration for an electron; these two modes can be made to correspond with the two forms of O—H—O and F—H—F in Fig. 6. The resulting energy state is lower than that of either single mode and the phenomenon is known as *resonance*.

Van der Waals bonds

Quite apart from the chemical bonds just discussed, all molecules attract one another to a small extent, the force of attraction falling off very rapidly with distance. Such forces are said to be caused by induced charges from polar molecules or dipoles; although very weak compared with chemical bonds, they are strong enough to produce cohesion in liquids and are believed to play a part in the bonding of certain minerals.

Co-ordination number

Having discussed the various types of bonding in solids, we can now consider how a crystal lattice is built up. Although the bonds in inorganic crystals are not wholly ionic, there is sufficient ionic character for us to regard the constituent atoms as ions, and we may picture the latter as charged spheres of a definite radius, called the ionic radius. In oxides and silicates, oxygen is the most electronegative element and so occurs as the ion O^{2-}, whilst the more electropositive elements, silicon, aluminium, etc., occur as Si^{4+} and Al^{3+}.

As previously mentioned, a lattice is a continuous, three-dimensional structure in which the valency forces or 'electron clouds' around each ion are 'shared out' between its neighbours. Thus, it is incorrect to single out a 'molecule' in a crystal of sodium chloride, since each Na^+ ion is effectively linked to six Cl^- ions which surround it, and vice versa. According to Pauling's rules, the valency forces of each ion are shared out between the ions surrounding it; thus, if the Si^{4+} ion is surrounded by four oxygen ions, as is generally the case, each oxygen receives one positive valency share from the silicon (Fig. 7). Conversely, each positive valency share from silicon is satisfied by one negative valency share from oxygen, leaving one

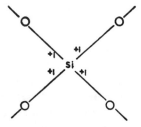

Fig. 7. The SiO_4 group.

Table 4

Ion	Radius (Å)	Radius ratio	Predicted co-ordination	Observed co-ordination
B^{3+}	0·22	0·16	3, (4)	3, (4)
Be^{2+}	0·34	0·24	4	4
Si^{4+}	0·39	0·28	4	4
Al^{3+}	0·57	0·41	4, 6	4, (5), 6
Mg^{2+}	0·78	0·56	6	4, 6
Na^+	0·98	0·70	6	6, (8)
Ti^{4+}	0·69	0·49	6	6
Zr^{4+}	0·87	0·62	6, (8)	6, (8)
Ca^{2+}	1·06	0·76	8, (6)	7, 8, 9
K^+	1·33	0·95	8, (9)	6, 7, 8, 9, 10, 12
Fe^{3+}	0·67	0·48	(4), 6	4, 6
Fe^{2+}	0·83	0·59	6	6

oxygen valency free to link to other positive ions. For the crystal as a whole to be electrically neutral, the number of positive charges must equal the number of negative charges, i.e. the number of positive valency shares associated with each ion must equal the number of negative shares.

The number of ions surrounding a given ion is called its *co-ordination number*; this latter is determined by the relative sizes of the different ions, on the assumption that they are packed as closely as possible. Thus, for a given ion of co-ordination number n, the n ions surrounding it are assumed to touch the central ion and one another, as far as possible to confer maximum stability. In the compound SiO_2, for example, each Si^{4+} ion of radius about 0·39 Å is surrounded by four O^{2-} ions, each of radius 1·40 Å, and it can be shown geometrically that this is the maximum possible number of surrounding ions if all are to be in contact. This maximum number, or co-ordination number, depends on the ratio: radius of cation/radius of anion. Thus, as the central cation increases in radius, the co-ordination number increases, as shown in Table 4. It would, of course, be possible to surround an oxygen ion with a very much larger number of silicon ions, but it would then be impossible to satisfy the valency requirements with this reverse arrangement. The ionic radii of a number of common cations, the radius ratio (with respect to oxygen) and the calculated and observed co-ordination numbers are shown in Table 5. The observed and calculated values agree quite well, but departures from the calculated values are to be expected because an oversimplified picture has

Table 5

Co-ordination number	Type of co-ordination	Range of radius ratio
8	Corners of cube	0·732–1·00
6	Corners of octahedron	0·414–0·732
4	Corners of tetrahedron	0·225–0·414
3	Corners of triangle	0·155–0·225
2	Linear	0–0·155

been presented: the ions are not hard spheres of constant radius and of course the bonding is not wholly ionic.

READING LIST

C. W. BUNN, *Chemical Crystallography*, Oxford University Press, 1961.
R. C. EVANS, *An Introduction to Crystal Chemistry*, Cambridge University Press, 1961.
K. M. MACKAY, and R. A. MACKAY, *Introduction to Modern Inorganic Chemistry*, International Textbook Company Limited, 1981.
A. F. WELLS, *Structural Inorganic Chemistry*, Oxford University Press, 1984.

Chapter 2

Silica

Silicon is one of the most abundant elements in the earth's crust, but it occurs chiefly in combination with oxygen as silica, SiO_2, and with oxygen and other elements as silicates. Silica is a *polymorphic* substance, capable of existing in several different forms, all having the same empirical formula but differing in the arrangement of the structural units.

The Si—O bond has sufficient ionic character to enable us to apply the basic concepts of crystal lattices, and we may therefore regard all forms of SiO_2 as being composed of Si^{4+} and O^{2-} ions, forming a close-packed array of spheres of fixed radius. The radius ratio of silicon to oxygen is 0·28 (Table 4), corresponding to a predicted co-ordination number of 4, which agrees with the observed value for the majority of crystalline and amorphous forms of silica. In general, therefore, each Si^{4+} ion in silica is surrounded by four oxygen ions. If we imagine the centres of the four oxygen ions to be joined by straight lines, as in Fig. 8, the resulting geometrical figure is known as a *tetrahedron*, having a triangular base and

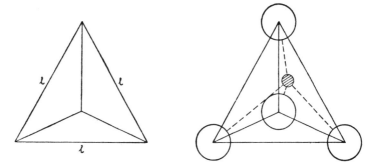

Fig. 8. The silicon–oxygen tetrahedron.

three triangular sides meeting at an apex; the Si^{4+} ion is situated at the geometrical centre of the tetrahedron, as shown. In silica, the tetrahedra are joined, corner to corner, to other similar tetrahedra, the various forms differing only in the way the tetrahedra are arranged.

QUARTZ

In quartz, the best-known crystalline form of silica, the Si—O—Si bonds joining neighbouring tetrahedra do not form a straight line but are bent round to give spiral chains (Fig. 9). Starting with any Si^{4+} ion, passing through silicon and oxygen alternately, spiral chains (Si_3O_3) can be traced throughout the structure, all in the same direction (in this instance, anti-clockwise). Every fourth silicon ion is a 'repeat' of the first one. The entire structure is built up by the linking of many such spiral chains through common silicon ions. It should be noted that in Fig. 9 and certain subsequent figures the ionic sizes are not drawn to scale, the distances between ions having been exaggerated for the sake of clarity. Special scale models of the forms of silica can of course be constructed and the reader is recommended to study these where available.

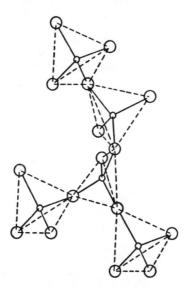

Fig. 9. Arrangement of silica tetrahedra in quartz (by courtesy of the British Ceramic Society).

STRUCTURES OF CRISTOBALITE AND TRIDYMITE

In both these structures the silica tetrahedra are linked to form rings, each containing six oxygen and six silicon ions (Fig. 10(a) and (b)). These Si_6O_6 rings are joined to form planes throughout the structures, each plane being linked to a neighbouring one by bridging oxygen ions. However, the

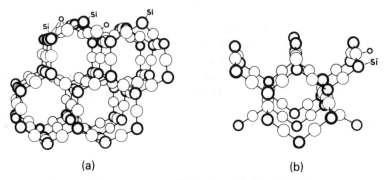

Fig. 10. The structures of (a) tridymite, (b) cristobalite.

six-membered rings are much more distorted in cristobalite than in tridymite; this is due to the different arrangement of the oxygen ions in the two structures, and is best appreciated by reference to Fig. 11(a) and (b), which is effectively part of a scale model, showing the oxygens only. In Fig. 11(a), two silicon–oxygen tetrahedra, as they are linked in tridymite, are depicted. Note that the lowest three oxygens form the triangular base of the lower tetrahedron; the second tetrahedron is inverted, with its base uppermost, and is joined to the lower one through a common oxygen ion which forms the apex of each. It will be clear that the three basal oxygens of the upper tetrahedron fall directly below corresponding oxygens in the base of the lower tetrahedron; this arrangement is repeated throughout the structure and corresponds to *hexagonal close packing* of oxygens, being characteristic of tridymite.

Comparing this with Fig. 11(b), which shows the oxygen arrangement in cristobalite, it is clear that the basal oxygens of the upper and lower silicon tetrahedra no longer correspond. Imaginary perpendiculars drawn from each of the upper three basal oxygens would fall midway between the oxygens of the lower tetrahedron, as if one of the tetrahedra in Fig. 11(a) had been rotated through approximately 60°. This arrangement of oxygens corresponds to *cubic close packing* and is characteristic of cristobalite.

In contrast to quartz, the symmetrical forms (β-forms) of tridymite and cristobalite have an Si—O—Si bond angle of very nearly 180°.

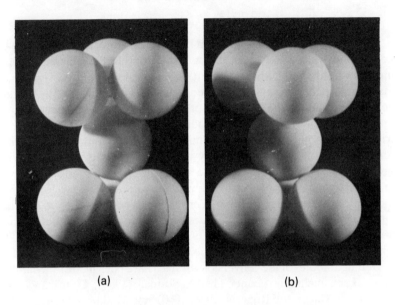

Fig. 11. Packing of oxygen atoms in (a) tridymite, (b) cristobalite (by courtesy of the British Ceramic Society).

CONVERSIONS

The three major crystalline forms of silica just described differ radically in structure, and although it is quite practicable to prepare cristobalite and (indirectly) tridymite from quartz, conversion is sluggish and involves breaking of silicon–oxygen bonds.

If quartz is heated above 1470°C for a considerable time, it is converted to cristobalite, probably via an intermediate amorphous phase. Cristobalite, in turn, if heated within the temperature range 870–1470°C, is gradually converted to tridymite. Both high-temperature forms are relatively stable when cooled rapidly down to room temperature and are not readily reconverted to quartz, except under hydrothermal treatment. These conversions may be summarised as follows:

$$\text{quartz} \xrightarrow{1470°C} \text{transition phase} \xrightarrow{1470°C} \text{cristobalite}$$
$$\text{(amorphous)}$$
$$\xrightarrow{870-1470°C} \text{tridymite}$$

It is uncertain whether quartz can be converted directly to tridymite by heat treatment alone. Where direct conversion has been claimed, some impurity has generally been present (often termed a *mineraliser*) which may stabilise the tridymite structure; in particular, sodium and potassium generally promote the formation of tridymite. In this connection it is interesting to note that pottery bodies fired at about 1200°C, which is within the stability range of tridymite, frequently contain cristobalite, formed from the quartz originally present in the raw materials. This probably means that the cristobalite has formed as an intermediate phase in the conversion of quartz to tridymite, the reaction therefore being incomplete at this stage of firing. Certain ions—notably calcium, barium and magnesium—accelerate the formation of cristobalite, although they are not essential to its formation. For this reason, lime has been added to silica bricks to promote the conversion of the raw quartz to cristobalite.

Detailed studies of the conversions of silica have been made by a number of workers, including Fenner (1919), Sosman (1965), Grimshaw *et al.* (1948), Chaklader and Roberts (1958, 1960), and Madden (1966).

INVERSIONS OF THE FORMS OF SILICA

Although the principal three forms of silica are relatively stable within the temperature ranges mentioned, certain minor modifications do occur. Both quartz and cristobalite exist at room temperature in the so-called α-form, which is actually a distorted form of the structures shown in Figs 9 and 10(b) respectively. When quartz is heated to 573°C or over, it is converted to a β-form, which has the 'ideal' structure shown in Fig. 9. Similarly, if cristobalite is heated at 220–280°C, it is converted to the β-form:

$$\alpha\text{-quartz} \xrightleftharpoons{573°C} \beta\text{-quartz}$$
$$\alpha\text{-cristobalite} \xrightleftharpoons{220-280°C} \beta\text{-cristobalite}$$

As indicated, these reactions are completely reversible, the β-forms reverting to α-forms immediately on cooling below the inversion temperature. Moreover, the inversions involve only a slight change in the Si—O—Si bond angle, and are therefore extremely rapid. The change from α-quartz to β-quartz is shown diagrammatically in Fig. 12, where the Si—O—Si bond angle is seen to *increase* as α → β.

Whilst the inversion temperature of quartz does not appear to vary, that of cristobalite has been found to depend on its previous history, and may

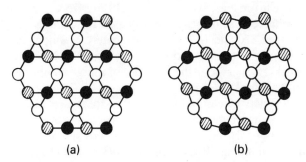

Fig. 12. Arrangement of silicon atoms in (a) β-quartz, (b) α-quartz (from *Structural Inorganic Chemistry*, by kind permission of A. F. Wells and the Clarendon Press, Oxford).

vary, as indicated, from 220 to 280°C. However, prolonged heating of cristobalite in general results in raising the inversion temperature up to a maximum of about 280°C, and this is accompanied by an increased regularity in the stacking of successive silicon–oxygen layers in the structure. Cristobalites with the lower inversion temperatures are therefore to be regarded as structurally disordered and become structurally more perfect with prolonged heat treatment:

$$\begin{array}{ccc}
\alpha\text{-cristobalite D} & & \beta\text{-cristobalite W} \\
\Big\updownarrow {\scriptstyle 220°C} & & \Big\updownarrow {\scriptstyle 280°C} \\
\beta\text{-cristobalite D} & \xrightarrow{1470°C} & \alpha\text{-cristobalite W} \\
\text{(disordered)} & & \text{(well crystallised)}
\end{array}$$

It may well be, as has been suggested, that the 'transition phase' previously reported during the quartz–cristobalite conversion may be identical with the cristobalite D above. The use of the notations 'D' and 'W' do not imply, of course, that there are only *two* well-defined varieties of cristobalite, but denote two extremes of a whole range of cristobalites having various degrees of disorder.

Like cristobalite, tridymite also undergoes a series of inversions, though it is by no means universally accepted as a pure silica phase. Some authorities maintain that there are two distinct varieties of tridymite, known as 'S' and 'M'; the latter exhibits two inversions:*

$$\text{tridymite M (I)} \underset{}{\overset{117°C}{\rightleftharpoons}} \text{tridymite M (II)} \underset{}{\overset{163°C}{\rightleftharpoons}} \text{tridymite M (III)}$$

the former five inversions, at 64, 117, 163, 210 and 475°C. Tridymite M, the less stable phase, is said to be converted to tridymite S on prolonged heating between 870 and 1470°C. Unfortunately, a wide variation in the inversion temperatures of various tridymites has been reported by different workers, suggesting the existence of a large number of tridymites differing in structural detail.

There is in fact increasing evidence that there are variations in the 'stacking' of the tridymite structural layers, occurring at regular intervals. Thus, the symbol 10H signifies a 'repeat unit' containing 10 layers, 20H a 'repeat unit' containing 20 layers. Both 10H and 20H types have been reported and may correspond to tridymites M and S above, but in view of the wide variations found in inversion temperatures, 'repeat units' of various magnitudes may exist in other tridymites.

OTHER FORMS OF SILICA

Silica gel

When solutions of sodium silicate are acidified, silica is precipitated as a gelatinous mass, which can then be extracted and dried. The resulting silica gel, as it is called, consists of silica tetrahedra joined in a random fashion, i.e. it is *amorphous*. The structure is highly porous and water and other liquids can be entrapped in the pores. In consequence, it has a high internal surface area—a property that is made use of in many ways. For example, because of its capacity of absorbing water, it is used as a drying agent.

When heated, silica gel evolves water continuously over a wide range of temperatures, indicating that some of the water is present not as molecular

---

*The α–β nomenclature is no longer used for all the various inversion forms of tridymite.

water, but as OH groups. The hypothetical 'silicic acid', Si(OH)$_4$, may be formed transiently in solution but has never been isolated.

Vitreous silica

When quartz, cristobalite or tridymite are heated to a temperature above 1710°C (the fusion point) and cooled rapidly, the silica tetrahedra are prevented from arranging themselves in a regular pattern and are constrained in the form of a glass, which like silica gel is also amorphous. Although formerly called 'silica glass', the term *vitreous silica* is now preferred for this form.

However, if vitreous silica is maintained at a temperature above 870°C for a long period, its recrystallises to cristobalite (and eventually to tridymite if conditions are suitable). This process is known as *devitrification*.

High-pressure forms of silica

Coesite is formed when silicic acid, flint or quartz are heated at 500–800°C under a pressure of 1.5×10^9 to 3.5×10^9 Pa.* When cooled, it is relatively stable at ordinary temperatures. The structure of coesite is said to consist of silica tetrahedra linked to form Si$_4$O$_4$ rings.

Keatite is formed by heating pure micro-amorphous silica (silicic acid) in the presence of water at 400–500°C under a pressure of 0.08×10^9 to 0.13×10^9 Pa. This form of silica is said to consist of a helical structure of Si$_4$O$_4$ rings.

Stishovite is produced by heating silica at pressures above 16×10^9 Pa at 1200–1400°C. Unlike the other forms of silica, stishovite has a structure like that of rutile (TiO$_2$) in which silicon is in six-co-ordination with oxygen; it is the only known compound of silicon containing six-co-ordinated silicon.

Silica W can be formed by cooling gaseous silicon monoxide, SiO, under carefully controlled conditions, when it decomposes to form silicon and silica W:

$$2SiO \longrightarrow Si + SiO_2$$

This form of silica is unusual in that it consists of silica tetrahedra joined at

---

*The pascal (Pa) is an SI unit of pressure, equal to 1 N m^{-2}.

edges rather than at corners, forming a fibrous chain structure. In moist air, silica W quickly hydrates to form silica gel.

PHYSICAL PROPERTIES OF THE VARIOUS FORMS OF SILICA

Pure quartz occurs as transparent, hexagonal crystals, and at 20°C has a density of 2·65 g ml^{-1}. It is noteworthy for the size and perfection of its crystals (frequently several inches in diameter). Quartz is optically active, rotating the plane of polarised light to the right or the left, depending on the direction of the spiral chains in its structure.

Tridymite does not often occur naturally but is a common constituent of fired siliceous products; it is occasionally found in volcanic lava. Under the microscope, it appears in thin sections as wedge-shaped crystals. The atoms in tridymite are less densely packed than in quartz, hence the density of the former is lower: 2·27 g ml^{-1} at 20°C.

Cristobalite, like tridymite, rarely occurs naturally but is the commonest constituent of fired siliceous materials, in which it can be observed microscopically as a mass of small crystals. Again, the packing of cristobalite is less dense than that of quartz and therefore the density is lower—in this instance 2·33 g ml^{-1}.

All three forms of silica show marked changes in thermal expansion coefficient at the inversion temperatures, as indicated in Fig. 13. The changes at the quartz and cristobalite inversion temperatures are particularly striking, being a frequent cause of spalling in silica refractories. Since there are many different species of tridymite, the curve shown in Fig. 13 should be considered as representative of one particular sample of tridymite only; different samples would give somewhat different curves, depending on their precise nature.

Since the linear expansion of silica at 1000°C is about twice that of clay, the thermal expansion of a pottery body is often purposely increased by the addition of flint, which on being fired transforms to cristobalite. Of all the forms of silica referred to, vitreous silica, as shown, has the lowest expansion (about 0·05% linear between 20 and 1000°C) and on this account it withstands sudden changes of temperature without shattering. Vitreous silica is therefore useful for making laboratory ware such as crucibles and tubes.

The high-pressure forms of silica, owing to closer packing of oxygen

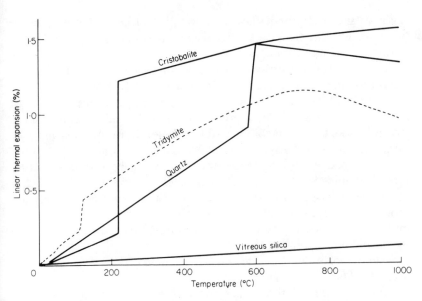

Fig. 13. Thermal expansion of the principal forms of silica.

ions, have densities higher than cristobalite or tridymite. Keatite, for instance, has a density of 2·50 g ml^{-1}, coesite 3·01 g ml^{-1}, and stishovite, the densest form known, 4·35 g ml^{-1}.

CHEMICAL PROPERTIES

With the exception of coesite, which is fairly resistant, crystalline and amorphous forms of silica are attacked by hydrofluoric acid, forming volatile silicon tetrafluoride:

$$4HF + SiO_2 = \uparrow SiF_4 + 2H_2O$$

This reaction is made use of in the gravimetric determination of silica. Most other common aqueous acids have no effect upon silica.

Although freshly precipitated silica and silica gel are attacked by aqueous NaOH, forming sodium silicate, the crystalline forms are more or less resistant. All forms of silica are, however, attacked by fused NaOH or fused sodium carbonate, the latter reagent being utilised in the decomposition of silicates for analysis.

The action of fused alkalis on silica is to form so-called soluble silicates, the precise nature of which is still in some doubt. Recent work by Dent Glasser (1982) and others indicates that at high pH, aqueous solutions of sodium silicate contain 'island' groups such as SiO_4^{4-} and condensed linear structures like $Si_2O_7^{6-}$ and $Si_3O_{10}^{8-}$. At low pH levels, various ring and cage forms of silicon and oxygen have been detected, in addition to those mentioned above. This strong tendency to form highly polymerised structures containing Si—O—Si bonds is characteristic of silicon; relatively few other oxides display the same tendency. These polymeric groups are probably the precursors of the final product of polymerisation, silica gel, mentioned previously; this product being formed when the pH is lowered by the addition of mineral acids.

Solutions of sodium silicate, known commercially as *water-glass*, are used as deflocculants, adhesives and for surface treatment of concrete.

Crystalline silica resists attack by slags rich in iron oxide and can therefore be used as a refractory material in steel-making, provided the alkali and lime content of the slag is low.

OCCURRENCE OF SILICA

Silica occurs naturally as *quartzite, ganister*, sand and sandstones, flint pebble, and as the semi-precious stones chalcedony, opal and agate. Quartzite rock occurs as a Carboniferous deposit (see Tables 8 and 9, pp. 50–51), forming part of the Millstone Grit, in South Wales and the Pennines; some quartzites are also to be found in some of the older (pre-Cambrian) rocks in Scotland. A typical Welsh quartzite contains some 97% SiO_2, the chief impurities being Al_2O_3 and Fe_2O_3, with an alkali content of less than 0·5%. Sandstones of suitable purity for refractories occur in parts of North and Central Wales, the Northern Pennines and the North Yorkshire Moors, but these have not been utilised so far.

Ganister, a Carboniferous deposit forming the seat-earth of coal seams, occurs in Derbyshire and Yorkshire, and also in the northern Pennines. It consists chiefly of very small particles of quartz, contaminated with a little clay, the latter providing a degree of plasticity which is of assistance in fabricating silica bricks. On average, a suitable ganister contains some 97% or more of SiO_2.

Flint is used as a source of silica where high purity is not so essential. A typical flint contains some 85% of SiO_2, the main impurity being calcium carbonate, but for use in whiteware the iron content should be low. Flint occurs in the form of nodules as a Cretaceous deposit in the Upper and Middle Chalk (see Table 8) and as pebbles on beaches adjacent to the chalk.

Flint consists of minute crystals of quartz (known as cryptocrystalline quartz) bound together by water molecules; in consequence, its density is less than that of massive quartz, being around 2·62 g ml^{-1}. When flint is calcined, the combined water is evolved at about 400°C and above. The bonding power of the water is thus lost and the flint pebble becomes friable and easily ground; finally, at about 1100°C, much of the quartz is converted to cristobalite, the latter conversion being promoted by the calcium oxide impurity.

Other sources of silica

Since the reserves of high-purity silica that are readily extracted economically in the United Kingdom are limited, attention has been turned to overseas sources.

One type of silica rock that has proved suitable as a refractory is *silcrete*, large deposits of which are found at Mossel Bay and Riversdale in the south-west of Cape Province, South Africa. Typically, this material contains some 97% of SiO_2, with 1–2% of TiO_2 and under 1% of Al_2O_3 and Fe_2O_3. Some quartzites are also to be found in northern Transvaal.

Another deposit of outstanding purity is the Findlings quartzite of northern Germany. In the United States, the principal sources of silica are the Alabama, Colorado and Wisconsin quartzites and the Pennsylvania sandstone. According to one US specification, the sum of the alumina, titania and alkalis in these deposits should not exceed 0·5%.

REFERENCES

CHAKLADER, A. C. D., and ROBERTS, A. L., *Trans. Brit. Ceram. Soc.*, **57** (1958) 115; **59** (1960) 323.
DENT GLASSER, L. S., *Chemistry in Britain*, **18** (1982) 33.
FENNER, C. N., *J. Soc. Glass Tech.*, **3** (1919) 116.
GRIMSHAW, R. W., WESTERMAN, A., and ROBERTS, A. L., *Trans. Brit. Ceram. Soc.*, **47** (1948) 269.
MADDEN, G. I., *Diss. Abstr.*, **26**(12) (1966) 7243.
SOSMAN, R. B., *The Phases of Silica*, Rutgers University Press, 1965.

READING LIST

J. H. CHESTERS, *Steelplant Refractories*, United Steel Companies Ltd, 1973.
R. W. FORD, *The Effect of Heat on Ceramics*, Institute of Ceramics Textbook Series, Elsevier Applied Science Publishers, 1967.
F. H. NORTON, *Refractories*, McGraw-Hill, 1968.
A. F. WELLS, *Structural Inorganic Chemistry*, Oxford University Press, 1984.

Chapter 3

Structure of the Main Types of Clay Minerals

GENERAL ASPECTS OF SILICATE STRUCTURES

Like the various forms of silica already described, all silicates, including the clays, are based on the $[SiO_4]^{4-}$ group. Before describing the clays in detail, it is helpful to glance at silicates in general and to note the various ways in which the silica tetrahedra are arranged in them.

The $[SiO_4]^{4-}$ group is of course incapable of independent existence and requires four positively charged units to balance the negative charges; these units may be other SiO groups, as in silica, or other cations, as in the silicates.

Island structures

In these structures, neutrality is achieved by metallic cations (principally Fe and Mg) being attached to the four oxygen ions of the $[SiO_4]^{4-}$ group. A good example of this type of structure are the olivines, which have the general formula $(Mg, Fe)_2SiO_4$. A complete crystal of olivine contains a very large number of these SiO_4 units with their appropriate cations; it forms a continuous whole because each atom is 'shared' by two or more other atoms.

Group structures

Instead of all the free 'oxygen valencies' being satisfied by metallic cations, two or more silica tetrahedra may be joined, corner to corner, to form a group structure (Fig. 14). Joining two groups together as in Fig. 14(a) gives the group Si_2O_7, with an oxygen 'bridge' connecting the silicon ions of

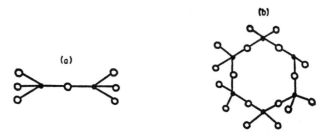

Fig. 14. Group structures.

adjacent tetrahedra. Again, six units can be linked to form a ring structure, as in Fig. 14(b), in which the basic unit is clearly $Si_6O_{18}^{12-}$.

Minerals of Si_2O_7 type are rare, but the group includes the mineral beryl or emerald, in which unsatisfied oxygens are linked to Al^{3+} and Be^{2+}, having the formula $(Be_3Al_2)Si_6O_{18}$. The mineral cordierite, $Al_3Mg_2(Si_5Al)O_{18}$, also has the ring structure. In this formula it should be noted that one of the six silicons in the ring has been replaced by an aluminium ion; this is an example of *isomorphous substitution*, where one cation can substitute for another of similar size without a radical change of structure. The substitution of trivalent Al for tetravalent Si upsets the charge balance; in this instance, neutrality is maintained by increasing the ratio of trivalent to divalent cations. Isomorphous substitution is a common occurrence in silicates and, as will be seen when discussing the clays, there are other possible ways in which neutrality can be achieved.

Chain structures

By joining an indefinitely large number of silica tetrahedra together, we arrive at a chain structure, which is common to two classes of mineral—the pyroxenes and the amphiboles (Figs 15 and 16).

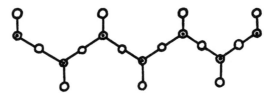

Fig. 15. The pyroxene chain.

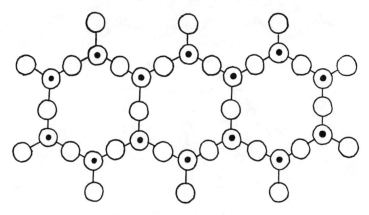

Fig. 16. The amphibole chain.

Owing to their chain structure, the pyroxenes and amphiboles are fibrous, asbestos-like minerals. In all these and subsequent structures, we can distinguish a 'repeat unit', i.e. the smallest group that can adequately represent the whole structure. In the pyroxenes, the repeat unit is SiO_3^{2-} and in the amphiboles it is $Si_4O_{11}^{6-}$.

Hornblende, a naturally occurring amphibole found in igneous rocks, has the general formula $(Ca, Na, K)_{2-3}(Mg, Fe, Al)_5(Si, Al)_2Si_6O_{22}(OH)_2$. As the formula implies, many varieties and degrees of substitution occur in this mineral.

Sheet structures

Carrying the notion of linking silica tetrahedra a stage further, it is not difficult to see that by condensing a number of pyroxene or amphibole units, a complete sheet of silicon–oxygen six-membered rings is formed (Fig. 17). Such sheets have a repeat formula $Si_2O_5^{2-}$ and are capable of indefinite extension in two dimensions; this is because the oxygen valencies within the plane itself are satisfied by being joined to two silicons, except for those at the boundaries, which of course are available for linking to other similar units. The only 'free' oxygen valencies are those at the apices of the silica tetrahedra, shown immediately above each silicon atom in the diagram; these oxygens cannot form sheets but may serve to link one sheet to another.

Other sheet structures, rather similar to the silicon–oxygen sheet, are

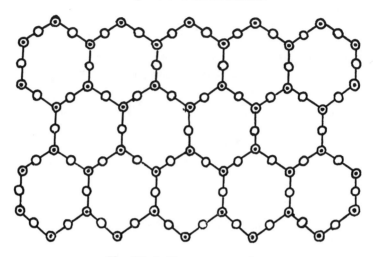

Fig. 17. A silicon–oxygen sheet.

present in the structures of gibbsite, $Al(OH)_3$, and brucite, $Mg(OH)_2$. Both these structures consist of sheets of Al or Mg atoms, and OH groups, the latter taking the place of oxygen in the silica sheets. Each Mg or Al atom is co-ordinated with six hydroxyls, each hydroxyl in turn being co-ordinated with two Mg or Al atoms. Many of the important clay minerals contain both silica and gibbsite or brucite sheets, as will be described shortly.

THE CLAY MINERALS

The kaolin-type unit

The *kaolin* group includes the clay minerals nacrite, dickite, kaolinite and halloysite. Their structures have one thing in common—they are composed of silica sheets linked to modified gibbsite sheets. Imagine a gibbsite sheet (Fig. 18(b)) placed directly over a silica sheet (Fig. 18(a)) in such a way that one in three of the OH groups is removed and replaced by the unsaturated 'vertical' oxygens of the silica sheet. These latter oxygens now form a 'bridge' between the two sheets, forming a composite unit layer of kaolin type. If we write the modified gibbsite layer as $[Al_2(OH)_4]^{2+}$ (i.e. having removed two OH groups) and the silica sheet as $(Si_2O_5)^{2-}$, we arrive at the composite formula $Si_2O_5 \cdot Al_2(OH)_4$, or $Al_2Si_2O_5(OH)_4$, the unit formula of the kaolin group.

The silica layer is often referred to as the *tetrahedral* layer, because of the tetrahedral shape of the SiO_4 groups. The co-ordination of the Si with respect to oxygen, which is four, is denoted by a Roman numeral immediately above the element thus: Si^{IV}. In the gibbsite layer, the oxygens are arranged so as to form the corners of octahedra, which are geometrical figures having eight faces and six corners. The co-ordination of Al is thus six and is represented as Al^{VI}. An important feature of the octahedral layer

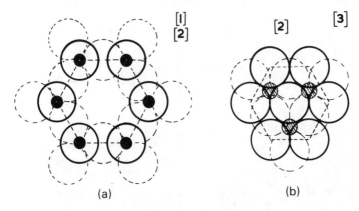

Fig. 18. (a) A silica sheet; (b) a gibbsite sheet (from *Structural Inorganic Chemistry*, by kind permission of A. F. Wells and the Clarendon Press, Oxford).

in the kaolins is that only two out of three possible sites are occupied by aluminium ions, the remainder being vacant; such structures are therefore called *dioctahedral*. There are three possible ways of filling three sites by two ions, thus giving rise to one source of variation in the unit layers.

The *serpentine* group of minerals have a structure consisting of a silica sheet condensed with a brucite sheet, with the unit formula $Mg_3Si_2O_5(OH)_4$. This group is clearly trioctahedral, this being possible because three atoms of divalent magnesium carry only the same charge as two atoms of trivalent aluminium. The best-known minerals of this group are *antigorite* and *chrysotile*, but these latter are not generally classed as clay minerals.

Stacking of layers

A crystal of a kaolin mineral consists not of one composite layer but of a very large number of such layers, which may be likened to a book, where each page represents a single layer (Fig. 19). Note that there is no ionic bonding between neighbouring units; the whole crystal is held together by hydrogen bonds, acting between OH groups of the gibbsite layers and oxygen atoms of adjacent silica layers. For this relatively weak secondary bonding to be effective, the appropriate oxygens and OH groups must be close together; there are several ways in which one unit can be 'stacked'

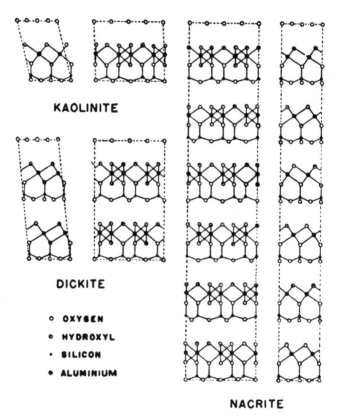

Fig. 19. Comparison of the structures of kaolinite, nacrite and dickite (by courtesy of G. W. Brindley and the Mineralogical Society).

upon another to achieve this bonding and this gives rise to four distinct minerals of kaolin type, viz. nacrite, dickite, kaolinite and halloysite.

In *nacrite*, the layers are stacked so that the atoms in one silica layer are directly above corresponding atoms in every other silica layer. Thus, the α- and β-angles (in addition to the γ-angle) are very nearly equal to 90°. Owing to the different ways of populating the octahedral positions, the structure only repeats after every sixth kaolin unit, making the c-dimension of the unit cell equal to 43 Å. The a-dimension is 5·15 Å and the b-dimension 8·96 Å, the unit cell as a whole being therefore practically orthorhombic.

In *dickite*, the unit layers are displaced regularly along the a-axis (and possibly the b-axis) so that the β-angle is no longer 90° but is equal to 96·8°. In the unit cell, $\alpha = \gamma = 90°$ and $a = 5\cdot15$ Å, $b = 8\cdot95$ Å and $c = 14\cdot4$ Å. There are thus two kaolin units in each unit cell, the latter being monoclinic.

In *kaolinite*, the units are again displaced regularly along the a-axis, so that although $\gamma = 90°$, the β-angle is 104·5° and the α-angle 91·8°. The unit cell is triclinic, with $a = 5\cdot15$ Å, $b = 8\cdot95$ Å and $c = 7\cdot39$ Å.

In *halloysite*, unit kaolin layers are still present but are displaced along both a- and b-axes in a random fashion, so that no values can be assigned to the α- and β-angles, though the γ-angle is still 90° and the unit cell dimensions for the *meta-form* (see below) are similar to those of kaolinite. Owing to the absence of hydrogen bonding between successive units, the structure is penetrable to water and a *hydrated form* exists, the formula of which may be written as $Al_2Si_2O_5(OH)_4 2H_2O$. This hydrated halloysite loses water readily at temperatures above 60°C and eventually is converted to the meta-form which has the same empirical formula as kaolinite. The extra $2H_2O$ in hydrated halloysite increases the c-dimension by about 2·9 Å.

Halloysite also differs from the other members of the kaolin group in its crystal habit. Whereas nacrite, dickite and (to a lesser extent) kaolinite exist as tabular or platy crystals, halloysite can exist in a tubular form, as has been shown by electron microscopy.

Disordered kaolinite

Although the principal clay mineral in the majority of British sedimentary clays is kaolinite, the latter appears to be less well crystallised than that of china clay. X-ray diffraction studies indicate a broadening of certain diffraction peaks (see Chapter 9), which has been interpreted by Plançon and Tchoubar (1977) as displacements in the vacant octahedral sites, of which two out of three only are occupied by aluminium ions. A less common cause can be the random displacement of unit layers along the b-axis, as in halloysite, or the curvature of layers, as observed in both halloysite and chrysotile.

In addition to the above, a distinct type of disorder is observed in many British sedimentary clays; where it has proved practicable to isolate the pure mineral, chemical analyses have shown that the composition differs significantly from that of 'ideal' kaolinite, there being in general a deficiency of aluminium, which has been isomorphously replaced by minor amounts of magnesium, iron and occasionally titanium. A typical ionic formula for such a kaolinite is:

$$(Al_{1.8}Fe_{0.1}Mg_{0.1})Si_{2.0}O_{5.0}(OH)_{4.0} \rightarrow Ca_{0.05}$$

The replacement of part of the trivalent Al by Mg results in an overall deficiency of positive charge, which is balanced externally by some other cation, commonly Ca^{2+}, which is then exchangeable for other cations. This substitution probably accounts for the high cation exchange capacity of disordered kaolinites occurring as fireclays, ball clays and some brick clays, as reported by Worrall *et al.* (1958). The degree of substitution appears to vary from clay to clay, the fireclays of the United Kingdom being apparently the most substituted. Between these and the well-crystallised kaolinites there exists a whole range of intermediate degrees of substitution; therefore it would not seem advisable to assign a specific name to these species but to use the non-committal term 'disordered kaolinite'.

At the present time it is not clear whether the chemical type of disorder just referred to is related in any way to the crystallographic disorder detected by X-ray diffraction, although the two often occur simultaneously, as for instance in certain British fireclays and ball clays. Since the substituent ions Mg and Fe are slightly larger than Al, it is reasonable to suppose that replacement of the latter by Mg and Fe could result in local distortions of the kaolinite structure. This would be compatible with the results of Plançon and Tchoubar, who showed by electron optical examination that a disordered kaolinite specimen contained folds, bubbles or inclusions at random points within the structure. From time to time, the possibility that disordered kaolinite could be an intimate association of 'normal' kaolinite and a small proportion of montmorillonite has been considered. Extensive chemical investigations by Sulaiman and Worrall (1982, 1985) have revealed no significant amount of montmorillonite, however; moreover, the high cation exchange capacity (some 30 meq $100\,g^{-1}$) would require the presence of 25–30% of montmorillonite, which would be readily detectable.

Attempts to quantify disorder or conversely the degree of crystallinity of kaolinites have been made from time to time using X-ray diffraction and, more recently, infra-red absorption spectrophotometry. Such attempts are complicated by there being several different types of disorder, perhaps two or more occurring simultaneously. The Hinckley method (Fig. 61, Chapter 9), based on the ratios of the heights of the $(1\bar{1}0)$ and $(11\bar{1})$ diffraction peaks, gives a crystallinity index that is useful for relatively well-crystallised kaolinites but is impracticable for more disordered types because the peaks given by the latter are too broad and ill-defined to measure accurately. For the latter, an index suggested by Parker (1969) may be more useful; this is based on the intensity I of the infra-red absorption peaks at $3700\,cm^{-1}$ and $3620\,cm^{-1}$. For clays contaminated with illite, Neal and Worrall (1977) found that the ratio: I_{3700}/I_{915} was preferable, since illite gives an infra-red peak near to the kaolinite peak at $3625\,cm^{-1}$ (Fig. 64, Chapter 9).

A significant feature of disordered kaolinites is that they are in general finer, of higher plasticity and higher unfired strength than well-crystallised kaolinites. There is, in fact, an inverse relationship between crystallinity index and plasticity, or unfired strength, as has been shown by Abboud (1978).

Anauxites

A mineral structurally similar to kaolinite but containing an excess of silica is known as anauxite. In the latter, the ratio of SiO_2 to Al_2O_3 may be as high as 3:1. It is difficult to imagine how the additional SiO_2 can be accommodated in the structure, but it has been suggested that double silica tetrahedral layers, presumably van der Waals bonded, are interleaved randomly between kaolin units. Although this still awaits confirmation, it is also possible that the anauxite is merely kaolinite contaminated with finely divided free silica, probably amorphous, from which it cannot easily be separated.

Allophane and imogolite

These aluminosilicate minerals, occurring in volcanic soils and sometimes as minor constituents of clays, are mainly amorphous, although imogolite shows a degree of one-dimensional order. The chemical composition of imogolite is said to approximate to $Al_2O_3.SiO_2.2\tfrac{1}{2}H_2O$, whilst that of allophane is more variable, the ratio of Al_2O_3 to SiO_2 ranging from 1·0 to 2·0.

Another distinguishing feature is the particle shape; under the electron microscope allophane is seen to consist of mainly spherical particles, whilst imogolite consists of long, thread-like particles.

According to Wada (1978), who has reviewed the nature, origin and properties of these minerals, both allophane and imogolite are soluble in 2% aqueous NaOH.

The montmorillonite group

We can carry the process of layer condensation a stage further by condensing *two* silica layers (one each side) with one gibbsite or one brucite layer, giving us respectively the minerals pyrophyllite, $Al_2Si_4O_{10}(OH)_2$, and talc, $Mg_3Si_4O_{10}(OH)_2$. Although these two substances are not true clay minerals, the montmorillonites may be considered as derived from them by the process of substitution. The structural features of the montmorillonites are shown diagrammatically in Fig. 20.

In these minerals, a single crystal is of course composed of a large number of units approximating to one or other of the above formulae, but since adjacent layers in these units are now silica layers only, there can be no outer hydroxyl bonds and the units are bonded by van der Waals forces. Such bonds are readily broken by shear and therefore the montmorillonites are easily cleaved and feel 'soapy' when rubbed between the fingers.

Although X-ray diffraction may enable a montmorillonite-type mineral to be recognised, it cannot always readily distinguish between members of the same group. The distinguishing features are in fact the nature of the substitutions and the latter are more easily determined by a chemical analysis of the purified mineral. According to Ross and Hendricks, whose views on montmorillonite structure are fairly widely accepted, the aluminium atoms in pyrophyllite are partly substituted by magnesium, iron or lithium; similarly, the silicon atoms may be partly substituted by aluminium, the only cation of similar radius. Similar considerations apply equally to talc, where the magnesium may be partly or wholly replaced by aluminium or iron.

Since the replacement of trivalent aluminium by divalent magnesium, atom for atom, results in an overall negative charge on the structure,

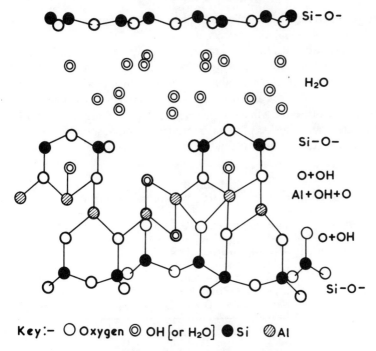

Fig. 20. The montmorillonite structure.

electrical neutrality has to be maintained by other cations external to the lattice; these cations are thus exchangeable. The very high cation exchange capacity of the montmorillonites (see later) is thus adequately explained by the high degree of substitution.

The ionic formulae of some representative minerals of the montmorillonite group are shown in Table 6.

It should be noted that the main structural units always carry a negative charge; even where an excess of positive charge exists in the octahedral layer, this is then more than compensated by a deficiency in the tetrahedral layer. Naturally occurring minerals often have formulae intermediate between any two of the 'ideal' formulae quoted above, and in fact an infinite variety of substitutions is theoretically possible. Naturally occurring montmorillonites of intermediate or mixed compositions may sometimes be considered as solid solutions of two or more 'ideal' compositions. Nontronite is of interest as it is ideally the completely iron-substituted member of the group. A further point of interest is that the magnesium-rich members of the group, hectorite and saponite, are clearly *trioctahedral*, the remainder *dioctahedral*.

Although in all cases the exchangeable cation is shown as sodium, this being the most commonly occurring cation, other species are encountered, notably calcium. The exchangeable cations are considered to be situated between the silica layers and because of the relatively weak bonding are able to diffuse outwards when the clay is in contact with water. In addition to this, water molecules occur in the interlayer position, some adsorbed by the charged surface and some held by the cations as water of hydration.

Table 6
Empirical Formulae of Some Montmorillonites

Name of mineral	Empirical formula	Exchangeable ions
Montmorillonite	$Al_{1.67}Mg_{0.33}Si_4O_{10}(OH)_2$	$Na_{0.33}$
Nontronite	$Fe_2Al_{0.33}Si_{3.67}O_{10}(OH)_2$	$Na_{0.33}$
Beidellite	$Al_2Si_{3.67}Al_{0.33}O_{10}(OH)_2$	$Na_{0.33}$
Hectorite	$Li_{0.33}Mg_{2.67}Si_4O_{10}(OH)_2$	$Na_{0.33}$
Saponite	$Mg_3Si_{3.67}Al_{0.33}O_{10}(OH)_2$	$Na_{0.33}$

This water, as discussed later, can be eliminated at a comparatively low temperature (150–300°C).

It is generally accepted that the unit layers in the montmorillonites are superposed in a completely random manner (a turbostratic type of disorder) due to the relatively weak bonding between the layers. Because of the great variety of substitutions possible, the unit cell dimensions vary from type to type, but are approximately: $a = 5.3$ Å, $b = 9.2$ Å, $\beta = 97°$, the value of c being variable. As described later, the unit layers can be forced apart by water and other polar liquids, causing the basal spacing (the perpendicular distance between equivalent planes) to vary from about 10 to 15 Å for water.

The foregoing general description of the structure of the montmorillonites is based on that of Hofmann, Endell and Wilm, as modified by Marshall, Maegdefrau and Hendricks. Some authorities, notably Edelman and Favejee, have suggested a somewhat different structure. They assert that some OH groups are attached to silicon, the structure being balanced by 'inverting' some tetrahedra and replacing some hydroxyl by oxygen in octahedral layers. The formula thus obtained is given as:

$$Al_2Si_4O_{10}(OH)_2 . n . H_2O$$

in which there are no substitutions, the cation exchange capacity being supposedly due to exchange of hydrogen in hydroxyl groups of silica layers. The rare occurrence of the Si—OH group in silicate minerals and the variable nature of OH ionisation make this hypothesis unlikely. If ionisation of hydrogen from hydroxyl groups were the sole cause of cation exchange in montmorillonites, the exchange capacity would vary with pH, yet there is no evidence that it does so.

Some confusion has arisen from time to time through the use of the word 'montmorillonite' to indicate both a specific mineral and the group name. For this reason, various alternative group names have been proposed, the one currently favoured in the U.K. being *smectite*.

The mica group

Only the very fine-grained micas may be thought of as clay minerals, but their structures are best understood by considering first the large-grained varieties.

In montmorillonites the degree of substitution rarely exceeds one-third of an atom per unit formula in any one layer. Imagine that, starting with pyrophyllite, $Al_2Si_4O_{10}(OH)_2$, we substitute one whole atom of Al for a Si atom in the tetrahedral layer, to form the negatively charged unit $Al_2(Si_3Al)O_{10}(OH)_2^{2-}$. Let the charge deficiency now be balanced by one atom of potassium, and we arrrive at the formula of *potash mica* or *muscovite*, $K.Al_2(Si_3Al)O_{10}(OH)_2$ or, as it is frequently written, $KAl_3Si_3O_{10}(OH)_2$. The first formula is more informative, since it shows that one of the aluminium atoms substitutes for silicon in the tetrahedral layer, the other two aluminium atoms forming part of the octahedral or gibbsite layer.

The essential difference between this structure and a montmorillonite is that in the mica there is a comparatively big charge, *concentrated in one layer*, so that the balancing cation, potassium, is very strongly held and therefore not exchangeable. Another factor contributing to the 'fixation' of the potassium is the fact that it fits closely between adjacent silicon–oxygen hexagonal rings, thus acquiring the twelve-co-ordination with oxygen which is required for closest packing (Fig. 21).

Similarly, if we start with talc instead of pyrophyllite, and perform (theoretically) the same substitution as before, we arrive at the formula of *phlogopite*, the trioctahedral magnesium analogue of muscovite:

$$K.Mg_3(Si_3Al)O_{10}(OH)_2$$

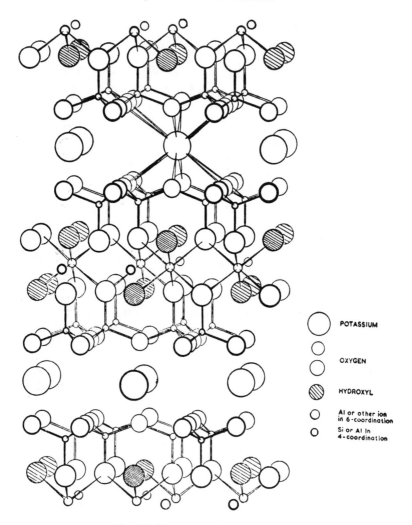

Fig. 21. The muscovite structure.

A sodium mica, analogous to muscovite, also exists and has been given the name *paragonite*. The calcium mica *margarite* is based on a similar type of structural unit but contains two substituent aluminium atoms:

$$Ca . Al_2(Si_2Al_2)O_{10}(OH)_2$$

The lithium-containing mica lepidolite is interesting in having the substituent atoms in the octahedral, rather than the tetrahedral, layer. A wide range of other substitutions occurs; for example, biotite, in which a considerable replacement of aluminium by iron has occurred. The formulae and mineralogical names of the more important micas are summarised in Table 7.

It should be noted that the forces bonding the balancing cations in the micas, although stronger than in the montmorillonites, are still weaker than those bonding atoms within the layers; hence the micas tend to be easily cleaved in a direction parallel to the sheets.

Table 7
Formulae of Various Micas

Muscovite	$KAl_2(Si_3Al)O_{10}(OH)_2$	Dioctahedral
Paragonite	$NaAl_2(Si_3Al)O_{10}(OH)_2$	Dioctahedral
Phlogopite	$KMg_3(Si_3Al)O_{10}(OH)_2$	Trioctahedral
Margarite	$CaAl_2(Si_2Al_2)O_{10}(OH)_2$	Dioctahedral
Biotite	$K(Mg, Fe)_3(Si_3Al)O_{10}(OH)_2$	Trioctahedral
Lepidolite	$K(AlLi_2)Si_4O_{10}(OH)_2$	Trioctahedral

As with the kaolins, variations in the stacking of successive mica units can occur, giving rise to *polymorphism*. Disordered structures are also known to exist, particularly in the fine-grained micas that occur in clays.

The illite group

Many natural clays contain a micaceous mineral, resembling muscovite in some respects, but containing less potash and more combined water than the normal muscovite formula allows. It is a fine-grained material, occurring frequently in sedimentary clays and often associated with montmorillonite or kaolinite. It is by no means certain whether this micaceous material is a single mineral or a mixture, but it is convenient to use the term *illite* for this material, on the understanding that it is not meant to imply a definite mineral of fixed composition but a group of minerals. Since the term is fairly generally accepted, the various other names previously suggested for mica-like clay minerals, such as hydrous mica, hydromica, sericite and bravaisite, may well be dispensed with.

Composition and structure of illites

Although not all reported illites are necessarily pure, analyses from various localities show potash contents ranging from 3 to 7%, silica from 38 to 53% and alumina from 9 to 32%. This departure from the normal muscovite composition could be explained in various ways. Some authorities, for instance, have suggested that the illites have less substitution of aluminium for silicon in the tetrahedral layer; this would account for the lower potash and higher silica contents often reported, but not for the higher combined water content.

Another possible explanation is that a proportion of the potassium atoms are replaced by hydroxonium ions (OH_3^+). This would account for both the low potash and high combined water values but not the higher silica contents. Even so, one would expect the OH_3 ions to be driven off at a different temperature from the hydroxyl groups when the clay is heated, but this does not appear to be the case. The thermal properties of illites show in fact a close resemblance to those of kaolinite, as described later.

There is a considerable body of evidence that illite constitutes one stage in the progressive breakdown of feldspar to disordered kaolinite; this hypothesis has been reaffirmed in recent work by Exley (1976). Recent research by the author and his associates suggests that the illites of British sedimentary clays consist of a 'core' of muscovite-type mica, closely associated with a breakdown product or transition phase that has some of the characteristics of kaolinite. The transition phase appears to be bonded to the unaltered material in such a way that the two cannot be separated by physical methods. Thus the physical properties of the composite material can be expected to be intermediate between those of kaolinite and of muscovite. If the above hypothesis is correct, it would clearly explain the apparent success of the rational analysis method calculations in predicting the physical properties of clays from their chemical compositions.

The chlorite group

Minerals of the chlorite group are best considered as derived from talc, as follows. Commencing with the 'ideal' talc formula, $Mg_3Si_4O_{10}(OH)_2$, suppose that one in every four silicon atoms is replaced by aluminium, resulting in the negatively charged structure $Mg_3(Si_3Al)O_{10}(OH)_2^-$, as in muscovite. Instead of balancing this negative charge with a single ion, however, imagine that neutrality is achieved by the insertion of a positively charged brucite layer, in which one magnesium atom is replaced by aluminium:

$$[Mg_2Al(OH)_6]^+ \quad \text{charged brucite layer}$$

Combining the charged mica unit with the charged brucite unit then results in the 'ideal' chlorite structure, which is of course electrically neutral:

$$\underset{\substack{\text{brucite}\\\text{layer}}}{Mg_2Al(OH)_6}\underset{\substack{\text{talc}\\\text{layer}}}{Mg_3(Si_3Al)O_{10}(OH)_2}$$

or, as the formula is usually written, $Mg_5Al_2Si_3O_{10}(OH)_8$. Naturally occurring chlorites may deviate considerably from the above formula because of further possible substitutions of Al for Si in the tetrahedral layer and of Fe^{2+}, Fe^{3+} for Al in the octahedral layer. In a crystal of chlorite, the sequence talc–brucite is repeated many times.

Chlorite is said to originate as either a primary mineral in igneous rocks or as a secondary product formed by the breakdown of biotite, hornblende and other minerals. There is also some evidence that it is an intermediate product in the breakdown of muscovite or illite to form kaolinite. Such a product would clearly need to be an aluminous chlorite, with alternate gibbsite and pyrophyllite-type layers, and would have a chemical composition identical to that of kaolinite. No such aluminous chlorite is known, although chlorites have been reported in which there is considerable substitution of magnesium by aluminium. Chlorite was found by the author to be one of the major constituents of a Midland brick clay, from which kaolinite appeared to be completely absent.

The palygorskite group

The mineral palygorskite is of little importance to the British ceramic industry but is of interest in connection with oil drilling.

The basic structure consists of double silicon–oxygen chains, similar to those in the amphiboles, parallel to the c-axis, joined together at their ends by oxygen ions. At intervals, there are magnesium– or aluminium–oxygen–hydroxyl units, cross-linking the silica chains. Between the silica chains there are gaps forming channels which are occupied by chains of water molecules also running parallel to the c-axis.

The ideal palygorskite formula is:

$$(OH)_2 . Mg_5Si_8O_{20} . 8H_2O$$

in which there may be some substitution of magnesium by aluminium. Four of the water molecules in the formula are believed to be situated in the structural channels. The name 'attapulgite' has been given to a mineral of this group in which the Al/Mg ratio is about 1:1. The palygorskites are fibrous minerals, owing to the easy cleavage between the chains.

A peculiarity of the structure is that four water molecules, presumably those situated in the channels, are driven off when the mineral is heated up to 400°C. The remaining four water molecules are more strongly bonded to octahedral aluminium or magnesium and are thus retained up to a higher temperature.

The mineral *sepiolite* is similar in chemical composition and physical properties to the palygorskites and its structure differs only in detail from the latter.

Vermiculites

The mineral vermiculite is structurally related to the micas and may be regarded as phlogopite or biotite, in which the potassium has been replaced by hydrated magnesium ions. The amount of net substitution is, however, rather less than in the micas and a typical formula may be written:

$$Mg_3(Si_{3\cdot 30}Al_{0\cdot 70})O_{10}(OH)_2 \rightarrow Mg_{0\cdot 35}4\cdot 5H_2O$$

The 4·5 molecules of water are believed to be co-ordinated with the magnesium and with the latter occupy roughly the same space as a brucite layer. That some of the structural water is not present as hydroxyl can be deduced from dehydration data, which show that interlayer water is lost in stages as the mineral is heated at temperatures up to about 500°C, but the remaining 'water', presumably derived from the hydroxyl groups, is not eliminated until a temperature of 700°C is attained.

As with most naturally occurring minerals, considerable deviations from the formula quoted above can occur, because of the variety and magnitude of substitutions that can occur. A characteristic feature of vermiculites is that if they are heated suddenly at about 500°C the interlayer water is evolved with such vigour that the layers are forced apart, causing the particles to expand to twenty or thirty times their original size, a phenomenon known as *exfoliation*. Vermiculites so treated are obviously highly porous and have been utilised as thermal insulators.

The interlayer magnesium in vermiculites is exchangeable for other cations. In natural specimens, some of the magnesium is found to be replaced partially by calcium.

FRAMEWORK STRUCTURES

Up to the present, structures that are capable of indefinite extension in two dimensions have been considered, namely, chain structures and sheet structures. Commencing as previously with the basic unit of structure, the SiO_4 group, it is possible to join these units together to form a framework structure which is capable of indefinite extension in three dimensions.

From the clay mineralogist's point of view, the most important framework structures are the feldspars, since it is from the latter that clays are believed to have been formed. The principal feldspars are the following:

Albite	$Na(Si_3Al)O_8$	Orthoclase	$K(Si_3Al)O_8$
Anorthite	$Ca(Si_2Al_2)O_8$	Celsian	$Ba(Si_2Al_2)O_8$

Essentially, the structures of all the feldspars consist of four-membered rings of type M_4O_4, where M represents Si or Al, linked laterally to similar rings and cross-linked with others to form a continuous three-dimensional structure, as illustrated in Fig. 22. The framework contains large voids, in which the cations Na, K, Ca or Ba are situated. In the structures of albite and orthoclase, the rings are of composition Si_3AlO_4, implying that one in

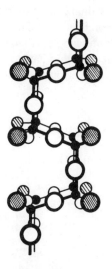

Fig. 22. The structure of feldspar (from *Structural Inorganic Chemistry*, by kind permission of A. F. Wells and the Clarendon Press, Oxford).

every four silicon atoms has been replaced by aluminium, resulting in an overall net charge on the framework which is balanced by the inclusion of sodium or potassium. In anorthite and celsian the substitution is two in every four, requiring the inclusion of a divalent ion to achieve neutrality. As with all naturally occurring minerals, many isomorphous substitutions are possible in the feldspars. Although feldspars are regarded as the principal parent mineral of clays, it is perhaps surprising that the former have not been found in ball clays and fireclays. They do, however, occur as secondary crystals in fired clay products.

MIXED-LAYER STRUCTURES

In discussing the montmorillonites, it was pointed out that the comparatively weak van der Waals bonds serve to hold the silica layers of successive units together, there being no possibility of hydroxyl bonding. These van der Waals type bonds can also link two or more different layers as a *mixed-layer* or *interstratified structure*. Thus, alternate layers of chlorite type and vermiculite type can be linked to form the regular sequence: CVCVCV . . . In such instances the crystal spacings observed by X-ray diffraction are directly related to the relative proportions of the two constituents and to their individual crystal spacings.

Where the interstratification is not regular, no such simple relationship exists and the problem of identification is then a very difficult one. Among the various mixed-layer structures that have been positively identified are illite–montmorillonite, chlorite–vermiculite and chlorite–illite.

REFERENCES

ABBOUD, D. Y., PhD thesis, Ceramics Department, University of Leeds, 1978.
EXLEY, C., *Clay Min.*, **11** (1976) 51.
NEAL, M., and WORRALL, W. E., *Trans. Brit. Ceram. Soc.*, **76** (1977) 57.
PARKER, T. W., *Clay Min.*, **8** (1969) 135.
PLANÇON, A., and TCHOUBAR, C., *Clays Clay Min.*, **25** (1977) 436.
SULAIMAN, I. S. M., and WORRALL, W. E., *Trans. Brit. Ceram. Soc.*, **81** (1982) 114; **84**, 146 (1985).
WADA, K., *Clays and Clay Minerals of Japan*, Elsevier Science Publishers, 1978, pp. 147–87.
WORRALL, W. E., GRIMSHAW, R. W., and ROBERTS, A. L., *Trans. Brit. Ceram. Soc.*, **57** (1958) 363.

READING LIST

G. W. BRINDLEY, and G. BROWN, *Crystal Structure of Clay Minerals and Their X-ray Identification*, Mineralogical Society, 1980.
R. E. GRIM, *Clay Mineralogy*, McGraw-Hill, 1968.
A. F. WELLS, *Structural Inorganic Chemistry*, Oxford University Press, 1984.

Chapter 4

Geology of the Clays

Any naturally formed aggregate or mass of mineral matter constituting an essential and appreciable part of the Earth is known to the geologist as a rock, this broad definition including such diverse things as coal and clay. There are three main types of rock, *igneous*, *sedimentary* and *metamorphic*. Igneous rocks are formed by the solidification of molten material, called magma, from the hot interior of the earth. Owing to the high temperatures and continual disturbances in the earth's interior, molten material has been forced to the surface at various periods in the earth's history, and has solidified to form igneous rocks of various ages.

Sedimentary rocks are formed when igneous rocks are decomposed by various agencies, and transported away from their source. *Epigenic agencies*, which include running water, carbon dioxide, winds and glaciers, slowly decompose igneous rocks and break them into very small fragments, which are carried down to the river estuaries, lakes or seas, and deposited on the seabed. These deposits later form sedimentary rocks, of which clay is a typical example. The deposits do not consist only of clay, because some constituents of igneous rocks decompose more slowly than others and may form other types of deposit, lying on top of earlier ones. Other products of epigenic action are soluble, e.g. calcium and magnesium bicarbonates, colloidal silica, and sodium and potassium salts. As the water carrying these substances became saturated, precipitation occurred, or accumulation of skeletal debris (the material of which had been extracted by the organisms concerned from the sea-water), forming beds of limestone, magnesian limestone and siliceous rocks. As each layer became submerged under others, it was subjected to increasing pressure and so consolidated into a more or less hard mass of rock.

In general, the older sedimentary rocks lie deeper than the younger ones, as evidenced by the type of fossils found and by radioactive measurements. However, the regular sequence of deposits has in places been interrupted by earth movements which resulted in folding and upheaval of the deposits, frequently submerging land masses by the sea, or alternatively raising seabeds to produce dry land. New rivers were formed during the process, with the result that some deposits were completely removed by the flow of water.

Hypogenic action results from the eruption of hot gases and vapours from the interior of the earth, which, on contact with igneous rocks, decomposes them to form new products. An example of hypogenic action is the alteration of the granites of Cornwall to form china stone and china clay. In this instance the alteration products were not transported but remained in close association with the parent rock; they are not therefore sedimentary but *residual* deposits.

The sedimentary rocks are classified according to age; Table 8 shows the various geological eras and periods, their approximate age and the principal types of rock formed during each period.

COMPOSITION OF IGNEOUS ROCKS

The type of mineral formed when the molten material cooled to form igneous rocks depended on the initial composition, the temperature of solidification, and the pressure. Generally speaking, the proportion of basic constituents increases with depth below the earth's surface; consequently the igneous rock known as *granite* is considerably less basic than *basalt*, the latter having originated at a greater depth. Thus, granite, a typically acidic rock, contains a high proportion of the siliceous minerals, orthoclase and anorthite; basalt, a typically basic rock, contains little feldspar but has a high proportion of the basic magnesian minerals, hornblende, pyroxene and olivine (Table 9). Table 9 also indicates the probable decomposition products formed when these two representative igneous rocks break down as already described. Although clay is common to both, a greater variety of decomposition products is obtained from basalt than from granite.

COMPOSITION OF SEDIMENTARY ROCKS

From Tables 8 and 9 it is evident that clay minerals are one of the important constituents of sedimentary rocks. With repeated deposition and sub-

Table 8
The Geological Systems

Era	Period	Approximate age (millions of years)	Principal clay deposits	Other deposits
LOWER PALAEOZOIC (or PRIMARY)	CAMBRIAN	500		Sandstones, slates, siliceous grits
	ORDOVICIAN	410		Sandstones, slates
	SILURIAN	350		Slaty and calcareous shales, limestones, flagstones, sandstones
UPPER PALAEOZOIC (or PRIMARY)	DEVONIAN	325	Brick shales	Old Red Sandstone, and limestones
	CARBONIFEROUS	285	Fireclays, Etruria Marl	Mountain limestone, millstone grit, Lower, Middle and Upper Coal Measures
	PERMIAN	210	Red and White marls	Magnesian limestone, gypsum and sandstones
MESOZOIC (or SECONDARY)	TRIASSIC	170	Keuper Marl	Bunter sandstone, pebble beds, Keuper Sandstone, gypsum
	JURASSIC	145	Oxford clay, Fuller's earth	Limestone, gypsum, calcareous sands and grit, jet and lignite
	CRETACEOUS	120	Gault clay, Hastings clay, Weald clay	Upper and Lower Greensands, chalk, flints
CAINOZOIC (or TERTIARY)	EOCENE	60	Dorset ball clays, London clay, Reading Beds, Bracklesham Beds	Pebbles, sands, lignite
	OLIGOCENE	35		
	MIOCENE	20		
	PLIOCENE	8	Devon ball clays	Siliceous limestone, lignite
QUATERNARY	PLEISTOCENE	1	Alluvium brick-earths Glacial clays, Boulder clays	Calcareous sands, loams and limestone Gravel, chalk, flinty shingle, sands
	HOLOCENE (or RECENT)	—		Sand and gravel, alluvial and glacial drift

Table 9
The Composition of Igneous Rocks and their Breakdown Products

Mineral	Approximate formula	Approximate % present		Probable decomposition products
		Granite	Basalt	
Orthoclase	$KAlSi_3O_8$	70	10	Kaolinite, colloidal silica, K_2CO_3
Anorthite	$CaAl_2Si_2O_8$			Kaolinite, colloidal silica, $CaCO_3$
Quartz	SiO_2	25	—	Unchanged
Hornblende	$(Na, K)_2(Fe^{2+}, Mg)$ $(Fe^{3+}, Al)_4 Al_2Si_6O_{22}(OH)_2$	—	90	Kaolinite or montmorillonite, limonite, haematite, $CaCO_3$, $MgCO_3$, colloidal silica
Pyroxene	$(Mg, Fe)SiO_3$	—		Colloidal silica, limonite, haematite, $MgCO_3$
Olivine	$(Mg, Fe)_2SiO_4$	—		As for pyroxene
Muscovite	$KAl_3Si_3O_{10}(OH)_2$	5	—	Probably unchanged
Biotite	$K(Mg, Fe)_3Si_3AlO_{10}(OH)_2$		—	Kaolinite or montmorillonite, iron oxides, colloidal silica, $MgCO_3$, K_2CO_3

mergence the layers of deposited material were compacted under considerable pressure and became quite hard and 'rock-like'. If the pressure was high enough they may have formed *shale*, a hard, laminated clay rock; or by further action of heat and pressure, they may have been converted to *slate*.

The nature of the clay mineral formed depended on the climatic conditions prevailing at the time. In fairly warm, humid conditions, kaolinite appears to have been the predominant mineral, whereas in cool and somewhat drier conditions, montmorillonite predominated. The widespread occurrence of kaolinite-type clays in the British Isles indicates the vast changes in climate that have taken place since their deposition.

In addition to clays, other sedimentary rocks of widely differing composition were formed. Sparingly soluble salts such as calcium and magnesium carbonates and colloidal silica were precipitated from seawater and deposited on the seabed, forming limestone, magnesian limestone and silica rocks.

Deposits of organic matter have also been formed from time to time. The forests that flourished at various periods in the past gradually decayed and through a series of complex physical and chemical changes were converted into sedimentary deposits of lignite (an immature brown coal of Tertiary or Jurassic age), bituminous coal or anthracite, depending on physical conditions. These organic deposits are now found associated with clay seams—lignite with the ball clays, bituminous coal and anthracite with the fireclays. In some way, part of the organic matter became adsorbed on to the clay and cannot be separated from it very readily; this organic matter is very finely divided and has a marked influence on the physical properties of certain clays.

The effect of organic matter is so important that it warrants some further discussion. The woody tissue from which coal was eventually formed consists of *cellulose*, an aliphatic compound composed of a large number of anhydro-β-glucose units held together by acetal linkages (Fig. 23), the empirical formula being $(C_6H_{10}O_5)_n$, where n is equal to some 3000 units.

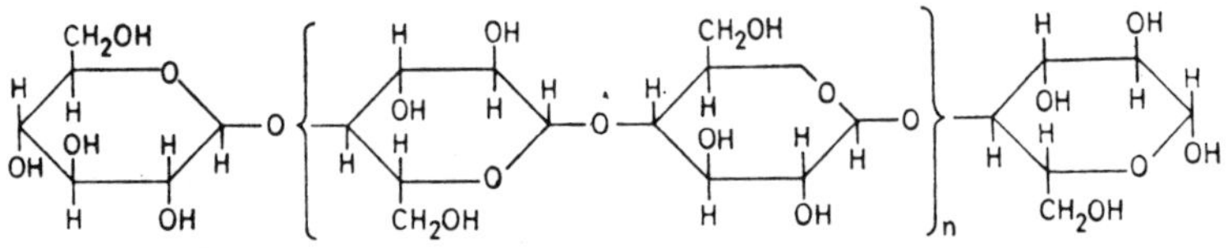

Fig. 23. The structure of cellulose.

In addition, a considerable proportion of a brown-coloured aromatic substance known as *lignin* is present; the formula of this substance is not known with certainty but it is believed to consist of a large number of units of the type shown in Fig. 24, having an empirical formula approximating to $C_{20}H_{28}O_7$, and joined together by ether–oxygen linkages. Certain minor constituents, notably waxes and resins, are also present. The former are essentially fatty acids and their esters, whilst the latter are condensed phenol-type compounds.

It is believed that in the transformation of wood to coal, the lignin side-chains become oxidised to carboxyl groups, forming the so-called

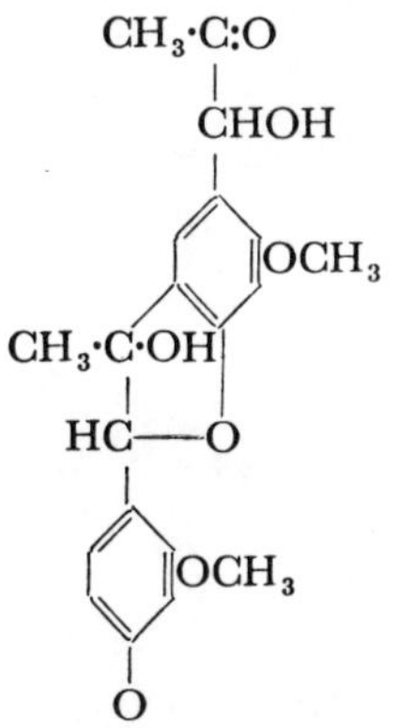

Fig. 24. The structure of lignin.

'humic acids'. The fate of the cellulose is uncertain, but some believe that it is completely decomposed to methane and carbon dioxide; others maintain that it is transformed, as is lignin, to humic acids. At this stage, the fossilised wood is an immature brown coal, often called lignite.

Older deposits, however, have undergone more drastic changes; intense heat and pressure due to further submergence and earth movements caused further condensation of the humic acids, producing condensed-ring compounds resembling graphite, except that some side-chains remain, as in Fig. 25. Such substances have been called *ulmins*. Simultaneously, the waxes and resins combined loosely with the ulmins, resulting in the familiar bituminous coal found in the Carboniferous system.

There is thus a marked contrast between the lignitious type of material associated with the younger clays and the bituminous coal associated with fireclays. Lignin and humic acids (lignin–humus) are highly colloidal materials, possessing cation exchange properties due to their functional groups —OH and —COOH. Bituminous coal, on the other hand, has very few functional groups and appears to have little effect on the colloidal properties of clays. Work by the author showed that the lignitious matter in ball clays could be divided into two distinct fractions: bituminous material, derived from waxes and resins, and capable of being extracted by organic solvents; and lignin–humus, which was insoluble but could be oxidised to carbon dioxide and water by hydrogen peroxide. The latter has a selective action and does not attack bitumen. Although the relative proportions of

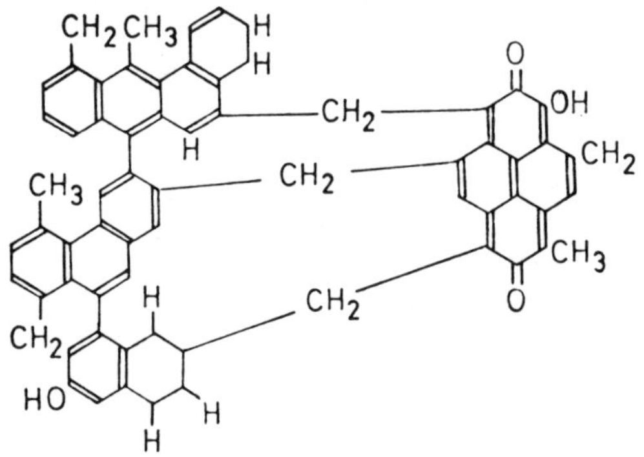

Fig. 25. Suggested structure of bituminous coal (by courtesy of the Institute of Fuel).

bitumen and lignin–humus were found to vary from clay to clay, the average composition of the organic matter associated with the ball clays was very similar to lignite.

The behaviour of the organic matter in fireclays is very different. Although a proportion of it can be extracted by solvents, the extract is chemically indistinguishable from the residue. Moreover, the action of hydrogen peroxide, although slow, is not selective but gradually oxidises the whole of the organic matter to carbon dioxide and water.

The decay of animal or plant remains produced deposits of oil and petroleum and is probably responsible for the accumulations of oil found in certain shales. Major deposits of oil and gas, including those under the North Sea, were probably formed in the same manner.

Metamorphic rocks have been formed by the action of heat and pressure on existing igneous or sedimentary rocks, altering their form but not

necessarily their chemical composition. A typical exa
version of shale to slate.

OCCURRENCE AND CLASSIFICATION OF C

The classification of clays has not followed a logical pattern but has depended partly on the use to which the clay is put, partly on general appearance, and partly on geology and location.

Two main types of clay, are, however, recognised, viz. *residual* and *sedimentary*, according to their geology. Residual clays, e.g. the china clays of Cornwall, are those clays that have not been transported by natural agencies and are to be found side by side with the altered igneous rocks from which they were formed. Sedimentary clays, by contrast, are those which have been removed from their origin by natural agencies.

Residual clays can usually be extracted from the parent rock and obtained in a comparatively pure state; on the other hand, sedimentary clays are rarely obtained pure, because many impurities are picked up and retained during transportation. The fine-grained nature of many such impurities makes them difficult or uneconomic to remove.

Residual clays

The only important deposits of residual clay in the United Kingdom are the china clays of Cornwall. Sometimes incorrectly termed 'kaolins' (a pharmaceutical term), they are the purest known form of the mineral

Table 10
Chemical Analysis of a Typical China Clay

Oxide	% by weight
SiO_2	46·29
Al_2O_3	38·38
Fe_2O_3	0·30
TiO_2	0·02
CaO	0·61
MgO	0·59
Na_2O	0·15
K_2O	0·51
Ignition loss at 1000°C	13·59

aolinite. The impurities total approximately 5–15% and consist chiefly of muscovite and paragonite with traces of the oxides of iron, calcium and magnesium. A typical analysis of a commercial china clay is shown in Table 10.

The location of the china clay deposits is indicated in Fig. 26, which may be compared with the simplified geological map, Fig. 27. China clay occurs in the western and central parts of the St Austell granite, the southwestern part of the Dartmoor granite and the western and southern parts of the Bodmin Moor granite. These granites were intruded into the earth's crust towards the end of the Carboniferous Period and were subsequently altered by both hypogenic and epigenic action during the Upper Cretaceous Period. The hypogenic agents responsible for the alteration of the granite were superheated steam and other hot gases containing highly reactive boron and fluorine compounds which, emitted from the interior of the earth, passed upwards through fissures in the granite, causing chemical decomposition. The epigenic agencies, mentioned earlier, were carbon dioxide and water.

According to Exley (1976) and others, it is now believed that the feldspars of the granite were first broken down to form mica or illite, the latter subsequently decomposing to form kaolinite or montmorillonite, according to prevailing conditions:

$$\text{Feldspar} \rightarrow \text{mica/illite} \begin{array}{c} \nearrow \text{kaolinite} \\ \searrow \text{montmorillonite} \end{array}$$

For the Cornish granites the conditions clearly favoured kaolinite but the reverse is true in certain other parts of the world. Again, whether disordered or well-crystallised kaolinite is formed (in this instance the latter) may well depend on the precise chemical composition of the feldspars and micas. From the work quoted above, it would appear that in the early stages of kaolinisation, soda feldspar is the more reactive, the potash feldspar being altered at a later stage. The kaolinisation process can nevertheless be represented by the general equation:

$$\underset{\text{feldspars}}{2(Na, K)AlSi_3O_8} + 3H_2O \rightarrow \underset{\text{kaolinite}}{Al_2Si_2O_5(OH)_4} + 4SiO_2 + 2(Na, K)OH$$

The characteristic way in which china clay occurs, i.e. as a fine-grained alteration product in granite rock, enables a special method of extraction to be employed. Large pits are dug in the altered rock to form a quarry; high-pressure jets of water are directed upon the sides, washing down the

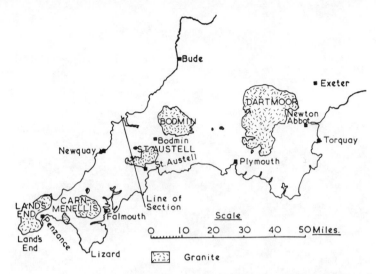

Fig. 26. The granite masses of south-western England (by courtesy of the British Ceramic Society).

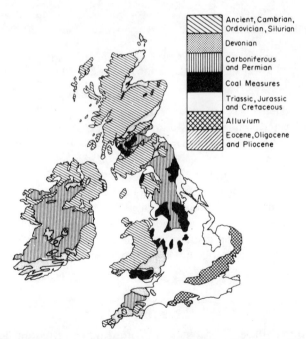

Fig. 27. Simplified geological map of the British Isles (from *Raw Materials*, by kind permission of Pergamon Press).

Fig. 28. A china clay pit near St Austell, Cornwall (photograph by courtesy of English Clays, Lovering Pochin & Co. Ltd).

fine clay and leaving behind much of the non-clay impurity. The clay–water slurry so formed is channelled into troughs where it is left for a short time to settle, thus allowing the coarse mica and quartz particles to settle out. The purified suspension is then pumped up to ground level into shallow tanks, where further sedimentation and grading is carried out (Figs 28 and 29). The excess water is then removed by a further period of sedimentation or by filter-pressing. Finally, the moist clay is dried in special kilns. China clay is now marketed in various grades according to use. Physical tests, including particle size distribution, wet-to-dry contraction, viscosity, casting rate, etc., are carried out at the producers' laboratories to check on the constancy of the product.

Composition

Owing to the efficiency of the method of extraction, English china clay is one of the purest sources of kaolinite. The content of silica and alumina is very close to that of the pure mineral (46·5% and 39·5% respectively) and

Fig. 29. A china clay works, showing settling tanks (photograph by courtesy of English Clays, Lovering Pochin & Co. Ltd).

the alkalis total less than 2%; the iron content (expressed as Fe_2O_3) lies between 0·5 and 1·2%. A calculation of the mineralogical composition indicates approximately 80–95% of kaolinite, with some 5–15% of mica. The remainder is mostly quartz and miscellaneous oxides, with a trace of montmorillonite. Organic matter is largely absent.

Particle size

The particle size range of china clay is more restricted than that of the sedimentary clays. The following data are quoted for the size distribution of a number of standard grades of china clay (Table 11).

Plasticity

Tables 11 and 12 show clearly that china clays, on the whole, contain less fine material than do ball clays; consequently they are less plastic than the latter and have less strength in the dry state. It follows that bodies

Table 11
Particle Size Distribution of China Clays

Type of clay	Particle size (%)		
	<1 μm	<2 μm	>10 μm
Porcelain clay	61	73	5
Bone china clay	36	47	13
Sanitary clay	20	31	23
Earthenware clay	30	43	24

containing only china clay and non-plastics are 'short' and difficult to manipulate. For this reason, plasticisers are now added to bone china bodies to improve their properties; the most common additives used are ball clay, bentonite and organic materials.

Cation exchange

China clays have a comparatively low cation exchange capacity, ranging from about 2 to 10 meq/100 g, the chief exchangeable ions being H, Ca, Mg, Na and K.

Deflocculation

Owing to their low cation exchange capacity, china clays require less deflocculant than sedimentary clays and are more sensitive to over-deflocculation. Sodium silicate alone does not completely deflocculate; polyphosphates are said to be more effective, but suspensions so treated are not very stable and moreover are likely to attack plaster moulds when used in casting.

Unfired strength

The unfired strength of china clays, after drying at 110°C, is lower than that of ball clays (see later), probably because the former are less fine-grained and contain little or no organic matter. Values from 0·4 to 2·7 MN m^{-2}* have been recorded.

*1 MN m^{-2} equals approximately 145 lb/in^2.

Colour

China clay is particularly valued for its whiteness, both in the raw and fired condition, due no doubt to the low iron oxide content. The whiteness of the raw material is particularly important for use in the paper industry and for this purpose the colour is checked periodically by appropriate instruments.

Firing shrinkage and vitrification

Values from 10 to 13% are quoted for the linear shrinkage of china clays fired at 1280°C. Because of the relatively low alkali content, china clay shows no appreciable vitrification at 1200°C.

Other sources of China clay

Comparatively few deposits in either Europe or America are as pure as the St Austell china clay. Deposits of a residual clay, similar to china clay, are found in central France. Near Echassières in the north-central portion is an important deposit of high purity which is used for the making of porcelain. In Germany, the Hirschaun deposit in Bavaria is one of the most important; good-quality kaolins are also located at Tirschenreuth and at Schneeberg and Seilitz in Saxony. A well-known deposit of high purity is found at Sedlec, near Karlovy Vary, Czechoslovakia.

The principal residual deposits of the United States lie in a 500-mile-long band, stretching from Vermont to Georgia and South Carolina and along the Mississippi valley. A few scattered residual deposits also occur in the west, the main ones being near Spruce Pine in North Carolina. The Georgia and Florida clays are of low iron oxide content and are suitable for paper-making and for whiteware. Unlike English china clay, they are of sedimentary (Cretaceous) origin. Few good-quality deposits of china clay type occur in Canada. There are some deposits in Quebec and northern Ontario but development has been hindered by the high cost of purification. Deposits of china clay type occur in Brakkloof and Nordhoek, near Capetown, South Africa.

Ball clays

The ball clays of the United Kingdom are sedimentary deposits that were laid down in the Eocene and Oligocene periods and were so called because they were originally dug out of the ground in blocks or 'balls'. They are an important raw material and are marketed both at home and abroad. There are in fact few other deposits of the same type and quality to be found elsewhere in the world. Careful examination of a large number of samples of ball clay has established that their composition varies widely, and therefore ball clays cannot be distinguished by composition but only by their location and geological age.

The areas in which ball clay is worked are comparatively small, and are confined to the valley of the Bovey and Teign, near Newton Abbot in

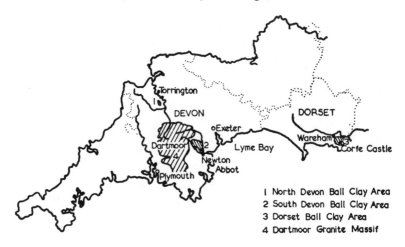

Fig. 30. The ball clay areas (by courtesy of the British Ceramic Society).

South Devon; a depression near Torrington in North Devon; near Wareham and Corfe Castle on the Isle of Purbeck; and also near Wimborne, Dorset (see Fig. 30).

A deposit somewhat similar to ball clays is the so-called pot-clay of Germany, such as Gross Allermarade and Klinginberg, which resembles siliceous ball clays. A deposit at Westerwald consists of a white-firing, fine-grained plastic clay, consisting mainly of disordered kaolinite and quartz, with some micaceous material.

The chief United States deposits of ball clay are to be found in Kentucky and Tennessee; there are also deposits in Florida, Alabama and New Jersey. These clays are of Lower Eocene origin and are sometimes adjacent to seams of lignite. Deposits of ball clay have recently been found in Mexico.

Geology of the Clays

In Canada, the only ball clay type deposit is the Whitemud Formation in southern Saskatchewan. These clays have been used in the making of pottery, stoneware and facing bricks. In South Africa, the nearest approach to true ball clays is found at Stellenbosch, near Capetown, and also near Albertinia, east of Capetown, and in the Witwatersrand (Transvaal).

Extraction of ball clays

The method adopted for extracting these sedimentary clays depends on the depth of the seam, the hardness of the clay, the depth of 'overburden' (material lying above it) and the inclination of the seam. *Deep mining* is usually economic only if the clay is very valuable and if some other mineral such as coal is being mined simultaneously. *Open-pit methods* are employed when the seam is not too deep.

Fig. 31. Underground ball clay working (photograph by courtesy of Watts, Blake, Bearne & Co. Ltd).

Fig. 32. Open-cast mining using rotary bucket excavator (photograph by courtesy of Watts, Blake, Bearne & Co. Ltd).

Ball clays are in fact mined both by open-pit and underground methods, according to their depth. For open-pit working, the overburden of soil is removed by mechanical excavators or bulldozers and the underlying clay is then dug out with pneumatic shovels and loaded on to lorries by means of an elevator (Figs 31 and 32). Where underground mining is employed, the workings seldom extend to more than 120 ft (36·5 m) below ground, the clay being hauled directly to the surface by mechanical excavator. No horizontal headings radiate from the main shaft, as they do in coal mining.

Chemical composition

The wide variations in ball clay composition are illustrated in Table 12. If the so-called siliceous ball clays are included, the maximum SiO_2 content rises to 80% and the minimum Al_2O_3 content falls to 15%. Outside these limits the deposits can no longer be regarded as clays and fall into the category of sands or loams.

The principal clay mineral of the ball clays is kaolinite, of various degrees of disorder, as described in Chapter 3. The major impurities are quartz and mica (possibly of illite type); minor impurities include organic

Table 12
The Variations of Chemical Composition of Ball Clays

Oxide	Range of variation (%)
SiO_2	40–60
Al_2O_3	25–40
Fe_2O_3	0·25–4·0
Na_2O	0–0·75
K_2O	0·5–4·0

matter (sometimes as much as 15%) and oxides of iron and titanium. Ball clays are variously described as 'blue', 'black' or 'ivory', according to their colour in the raw state; the black appearance is due to organic matter, whilst ivory clays owe their colour to iron oxide. The blue colour sometimes observed is probably due to scattering of light by the very fine particles of clay.

The organic matter of ball clays, as already mentioned, is of lignite type. Lignite has been analysed and has the following chemical composition (mineral-free basis):

Carbon	68·52%
Hydrogen	5·10%
Nitrogen	0·68%
Sulphur	4·67%
Oxygen	21·03%

Now by far the most accurately determined constituent of these is carbon, which can be determined by a modified combustion method, due allowance being made for any carbonate present. Assuming the organic matter in ball clays to have the same composition as that of lignite, we can readily calculate the total organic matter in a clay by multiplying the percentage of carbon by 100/68·52, or 1·46, the latter factor being the ratio of total organic matter to carbon.

It has already been pointed out that the organic matter of ball clays contains two principal constituents, bitumen and lignin–humus. It has been found possible to determine these substances separately and so obtain a direct measure of the total organic matter. These investigations indicate that the ratio of total organic matter to carbon varies slightly, depending on the relative proportions of the two constituents, but that 1·5 is a good average value, agreeing reasonably well with that obtained from the analysis of lignite. Thus, we have a rapid and convenient way of calculating organic matter content, which is of considerable assistance in the calculation of overall mineralogical composition. It is important to realise that the above considerations are valid only for ball clays, where the organic matter is of lignite type; for other clays the factor is obviously different.

Particle size distribution of ball clays

The average size distributions of ball clays from the three main areas are shown in Table 13. As is readily seen, a substantial proportion of all three clays consists of material less than $0.05\,\mu m$ radius. The material unaccounted for consists of material greater than $1\,\mu m$ and is obviously of little interest since it probably consists of quartz and other coarse-grained impurities.

Table 13
Average Particle Size Distribution of Ball Clays

Source	Material within the following size ranges (μm) (%)						
	<0.05	0.05–0.10	0.10–0.25	0.25–0.50	0.50–0.75	0.75–1.0	Total
North Devon	12.0	15.4	24.3	12.5	5.5	4.7	74.4
South Devon	9.6	15.7	29.2	17.9	5.8	3.7	81.9
Dorset	15.7	19.5	24.7	13.1	5.2	2.9	81.1

One aspect of the determination of particle size is worthy of mention here. For the clays, the method invariably used is that of sedimentation in aqueous suspension, assisted by centrifugation where particles of radius less than $1\,\mu m$ are present. For the test to be valid, the clay must first of all be thoroughly dispersed, usually by the aid of a chemical deflocculant. Since every grain of clay consists of a very large number of crystallites (called the 'ultimate particles'), a considerable time may be required before the agglomerated grains are broken down by the action of water. In the normal test, this breakdown is assisted by hand-shaking and the addition of 'Calgon' as dispersing agent. However, work by the author on fireclays has shown that more effective methods of dispersion, e.g. ultrasonic treatment, result in more effective breakdown and a higher recorded value for the submicron range. Since further examination confirmed that no crystallites were fractured by the process, it was clear that the increased yield of 'fines' was merely due to more complete breakdown of hard agglomerates. It should be added that fireclays are probably more difficult to break down than most other clays.

Specific surface area

Comparatively few data on the surface areas of ball clays seem to be available, but a reasonable value would seem to be around $20 \text{ m}^2 \text{ g}^{-1}$ for an average ball clay.

Wet-to-dry shrinkage

The wet-to-dry shrinkage of ball clays is comparatively high; in a survey of U.K. ball clays, values of up to 15% were quoted, with critical moisture contents from 10 to 32%. In practice, the non-plastic materials incorporated with ball clay in whiteware bodies reduce the shrinkage very considerably. The theory of shrinkage is discussed later, but it should be remarked that the highest values of wet-to-dry shrinkage are usually associated with a high proportion of fine particles.

Dry strength

The ball clays are notable for their high unfired strength, which is particularly useful in the manufacture of whitewares; in recent years it has become the practice to incorporate a small proportion of ball clay in china bodies to improve strength and plasticity. The high unfired strength of ball clays is due chiefly to the high proportion of fine particles and in many instances to the presence of organic matter. The effect of the latter can be enormous; experiments by the author showed that the addition of only 0·1% by weight of humic acid to a china clay increased its hardness by 100%.

Modulus of rupture

Values of up to $6·9 \text{ MN m}^{-2}$ have been recorded, the highest values being generally associated with the 'black' ball clays.

Cation exchange and deflocculation

The clay mineral of the majority of ball clays is the disordered form of kaolinite and has a high cation exchange capacity (c.e.c.), of the order of 30–40 meq/100 g. Since, however, non-clay impurities that do not contribute to cation exchange are almost invariably present, ball clays as mined have a c.e.c. somewhat lower than this, the average being around 15 meq/100 g. Because of the high c.e.c., ball clays respond well to deflocculants but the proportion required for complete deflocculation is rather higher than for china clays. Ball clays of high organic matter content can tolerate large excess additions of deflocculant. Work by the author shows that organic matter makes a large contribution to the c.e.c. of ball clays.

Soluble salts

Although occasional values approaching 1% have been reported, the average soluble salt content of the Devon and Dorset ball clays is 0·1–0·2% and is within normally accepted limits. Above about 0·5%, difficulties can arise in deflocculating clays for slip-casting. Most soluble salts found in the ball clays appear to be sulphates of calcium and magnesium, though traces of chloride do occur.

Fired colour

It is important in the manufacture of whiteware that the clays are 'white-firing' or nearly so, and most ball clays fulfil this requirement. The South Devon clays are reported to fire off-white to creamy-white, the North Devon clays pale ivory to ivory and the Dorset clays ivory, buff or red.

The principal factors affecting fired colour are the proportion of iron oxide, the state of oxidation of the iron after firing, the degree of subdivision of the oxide, and the extent of vitrification. Calcium oxide and magnesium oxide are said to bleach the reddish colour associated with iron oxide, whilst organic matter (e.g. in the 'black' ball clays) may lighten the colour by maintaining the iron in the ferrous state.

Vitrification

The extent to which a clay vitrifies at any given temperature is governed largely by the type and amount of fluxing oxides, namely the oxides of sodium, potassium, calcium and magnesium. The Dorset ball clays and certain South Devon clays vitrify appreciably at 1100°C; the North Devon and the remainder of the South Devon clays require a temperature of 1200°C before vitrification occurs.

Plasticity

The ball clays of the United Kingdom are the most plastic of the kaolinite-type clays. Although methods for measuring plasticity have been improved in recent years, most assessments in the past have been semi-quantitative and subjective, based on personal experience of working characteristics. This subject is discussed more fully in Chapter 7. In ball clays, as in the other sedimentary clays, the main plastic component is disordered kaolinite, the quartz and other accessory minerals contributing but little plasticity; there is evidence, however, that illite, where present, may contribute to plasticity. The organic matter of ball clays has been shown to increase plasticity enormously, the so-called black ball clays of high organic content being particularly plastic. A leading clay producer in the UK has developed a process for refining ball clay using a combination of wet and dry methods. By this means a clay free from harmful impurities and of reasonably constant composition is obtained.

The fireclays

The British fireclays are sedimentary clays that were laid down in the Carboniferous period, most of them associated with the Coal Measures or the Millstone Grit. The term 'fireclay' suggests the ability to withstand heat, and whilst a high proportion of the Coal Measure clays fall into this category, others are not true refractory clays but may be used for making sanitary ware, buff tiles, engineering bricks, etc. For high refractoriness, the alumina content should be high (approaching 30–40%) and the alkali content low (less than 1%). Fireclays have been subjected to considerable pressure because of the depth of burial, causing them to be hard and compacted. The so-called underclays, situated immediately beneath the coal, are said to be the most plastic fireclays.

Occurrence in the United Kingdom

Northumberland and Durham. Almost all the clays in these counties are extracted as by-products of coal mines or open-cast sites. Alumina contents vary from 28 to 36%, the Durham fireclays being slightly higher in alumina. These clays are used for making sanitary ware or refractories, according to the alumina content.

Cumbria. Fireclay deposits in Cumbria lie chiefly along the coast and under the sea, extending from south of Whitehaven to Wigton. The only important workings are at Lowca, five miles south of Workington, where clay from the Micklam seam, containing some 35% alumina, is mined and used for the manufacture of firebricks and casting-pit hollow ware for steelworks. A similar fireclay occurs at St Helens colliery, just north of Workington, but this is not at present being utilised.

Lancashire and part of Cheshire. The fireclays in these counties cover a triangular area corresponding to the coalfield, running from Colne in the north to Huyton in the south-west and to Stockport in the south-east. In addition there is a narrow area extending to Macclesfield and to Chapel-en-le-Frith. One of the most useful fireclays in this area is the underclay of the Mountain Mine coal, worked in the Blackburn–Burnley district. Other useful underclays are worked in the Wigan, Horwich and St Helens district, and near Stockport, Macclesfield, Oldham and Bolton.

North Staffordshire. Here the fireclays correspond with the North Staffordshire coalfield, extending from Congleton in the north to Longton in the south and bounded in the east at Oakamore and Madeley respectively. Many of these clays are not highly refractory, but some are used for making bricks and tiles.

South Staffordshire (West Midlands) and Worcestershire. The fireclay area again corresponds to the coalfield, which extends from Armitage in the north to Halesowen in the south, with Wolverhampton and Walsall on the western and eastern boundaries respectively. The best-known fireclays of this district are those of Stourbridge. The Old Mine clay, with an alumina content from 25 to 40%, was much valued as a refractory clay. This has now been worked out, but has been replaced by deeper seams called the New Mine clay with some 30–40% alumina, of which large reserves are available. The clays are used for making firebricks, sanitary ware and salt-glazed pipes, depending on their refractoriness.

Warwickshire. The most important fireclay-producing area is between Tamworth and Nuneaton (the so-called Nuneaton clay). The latter is not very refractory but is used for making building-bricks, salt-glazed pipes and chemical stoneware.

Derbyshire and Leicestershire. The fireclays in these counties fall into two groups: (1) near Swadlincote, Church Gresley and Woodville, and (2) east of the Erewash valley. These clays are less refractory than those of Stourbridge, since they contain some 4% of lime, magnesia and alkalis.

Yorkshire. Fireclays of widely differing composition occur in Yorkshire and are concentrated in the Leeds and Halifax districts. Many of them provide high-quality refractory bricks; others, of lower refractoriness, are used for making pipes and tiles. Somewhat less refractory fireclays of a more siliceous nature are obtained from the Huddersfield and Sheffield districts.

The pot clay seam in the Stannington area of South Yorkshire, a Carboniferous clay mined by opencast methods, is used for casting pit ware and for grog production.

Lincolnshire. The only important deposits in this county are those in the Stamford district.

Shropshire. The Shropshire coalfield extends from Dorrington, almost due south to Bewdley and Stockton (Worcs.), but the chief fireclay area within this is in the Coalbrookdale district. These fireclays are contaminated with nodules of ferrous carbonate but can nevertheless be used for the manufacture of wall and floor tiles.

Scotland. The Scottish fireclays, like the Coal Measures, are concentrated in the Forth–Clyde valley. Some of the most refractory clays are found in this area, notably the Glenboig and Bonnybridge fireclays, which have an alumina content of almost 40%. The upper fireclay seams in Ayrshire are exceptional in that they contain free alumina in the form of bauxite, which naturally affords a high refractoriness. All the fireclays mentioned above can be made into high-grade refractory products. Slightly less refractory clays, like those worked near Barrhead, are used for ordinary firebricks and sanitary ware. Fireclays also occur in various other parts of Scotland but they are less refractory and serve the local markets only.

North Wales. The only important deposits in North Wales are those at Mold, Ewloe, Buckley, Ruabon and Wrexham.

South Wales. The main fireclay workings in this area are in the vicinity of Neath, Swansea, Llanelli, Merthyr Tydfil and Pontypridd. Most of the Welsh fireclays are highly siliceous and of moderate refractoriness only.

Northern Ireland. Some fireclay deposits are worked in Co. Tyrone; they are mostly of inferior grade and unsuitable for refractory purposes, but some are said to be very suitable for making salt-glazed pipes.

Other sources of fireclay

True fireclays are not particularly widespread in Europe. In Germany, deposits occur at Limburg, on the River Lahn; in the Westerwald region, between the Rhine and the Lahn; and at Pfalz, Meissen and Leutendorf.

There are also deposits in Normandy (France), at Oberpfalz (Austria) and in Czechoslovakia.

Fireclays are abundant in the United States, some of the more important deposits occurring in Kentucky, Ohio, Missouri and Pennsylvania. There are relatively few good fireclays in Canada, the most noteworthy being the Cretaceous fireclays of the Whitemud Formation in Saskatchewan, the Cretaceous and Tertiary shales of Sumas Mountain, British Columbia, and the low-grade Cretaceous fireclays in Nova Scotia.

In South Africa, fireclay is found in the Witwatersrand area; a flint clay occurs in a region from Pretoria to Belfast (Transvaal). A variety of illitic–kaolinitic clays, used both by the pottery and refractories industries, is found near Grahamstown (eastern Cape Province).

Extraction of fireclays

Fireclays are extracted by open-pit methods where practicable, but are also deep-mined simultaneously with coal.

Composition of fireclays

Some idea of the composition of UK fireclays can be obtained from Table 14, which shows the ultimate analyses of a number of fireclays used in the sanitary and tile industries. A wide range of compositions is shown; for instance, for a total of 126 samples, the silica contents ranged from 44·5 to 81·4% and alumina from 11·9 to 37·8%. It should be noted that this survey did not, of course, include the more refractory clays used for the manufacture of firebricks, some of which have alumina contents of some 40%. The highest proportion of carbon (determined by combustion) in the whole series was 5·7%, i.e. much lower than that found in ball clays and brick clays. The carbon content is of interest because of the effect it may have on certain physical properties and on firing.

Rational analysis

The chief minerals in the fireclays are said to be a poorly crystallised (disordered) kaolinite, mica and quartz. Minor impurities such as organic matter, carbonates of calcium, magnesium and iron, pyrites, hydrated iron oxide and anatase (a form of TiO_2) are also frequently present. It is interesting to note that whilst carbonates are frequently present in the fireclays, they are apparently absent in the UK ball clays; no explanation for this has yet been offered.

The percentages of the principal minerals have been calculated from the chemical analysis, assuming 'ideal' compositions for the minerals; the values obtained constitute a so-called rational analysis (Table 14).

Particle size distribution

Particle size distributions for a number of fireclays are shown in Table 14. It is unfortunate that a more detailed distribution for the submicron range is not available, since this contains the colloidal fraction, the proportion of which determines many of the physical properties. One observation is worthy of mention, however; the clays from the north of the British Isles are on the whole coarser than those from further south. This accords well with the accepted view that the clays were deposited, during Carboniferous times, by running water, travelling from north to south, since under these conditions the coarser material would sediment first, i.e. in the north, the fine materials travelling further south before deposition.

Critical moisture content

Since the critical moisture content is a measure of the water remaining in the pores of a clay test-piece at the point where shrinkage just ceases (see later), it ought to be related to the degree of packing of the particles and therefore to the size distribution. From the figures quoted in Table 14, there is no obvious correlation with particle size as such and one can only conclude that other factors, such as the degree of flocculation, are more important. This latter is discussed more fully in Chapter 8.

Unfired strength

Reference to Table 14 shows that on average the dry strength of fireclays is intermediate between that of ball clay and china clay. Although the clay mineral of fireclays is essentially similar to that of ball clays, two factors may account for the relatively low strength of fireclays: (a) a higher proportion of quartz and other non-clay material; (b) the high degree of compaction of the grains in the fireclay mineral, as a result of which it may be difficult to break down to its ultimate particles. This situation would be encountered to an even greater extent with shales.

Table 14
Some Properties of Fireclays from Various Localities

Chemical analysis (%)	South Midlands		Yorkshire		Northumberland and Durham		Scotland		Staffordshire and Shropshire	
	Mean	Range	Mean	Range	Mean	Range	Mean	Range	Mean	Range
SiO_2	56.5	45.9–67.3	60.1	53.4–76.5	59.5	48.6–81.6	57.4	44.5–79.6	60.9	44.5–71.1
Al_2O_3	26.6	19.4–33.2	25.7	15.5–30.0	26.0	12.2–32.5	26.4	11.9–37.8	23.7	16.7–33.8
Fe_2O_3	2.4	1.3– 5.6	1.7	0.8– 2.8	1.6	0.6– 3.4	1.9	0.7– 4.2	2.2	1.1– 4.4
TiO_2	1.5	1.3– 2.1	1.2	1.0– 1.6	1.2	1.0– 1.6	1.2	0.7– 1.9	1.3	1.1– 1.5
CaO	0.3	0.1– 0.8	0.4	0.2– 0.5	0.4	0.1– 0.8	0.3	0.1– 0.6	0.3	0.0– 1.6
MgO	0.5	0.2– 0.7	0.5	0.2– 0.8	0.5	0.2– 0.8	0.5	0.2– 1.0	0.6	0.0– 1.7
Na_2O	0.2	0.1– 0.5	0.15	0.1– 0.3	0.1	0.0– 0.3	0.1	0.0– 0.4	0.2	0.0– 0.4
K_2O	1.4	0.4– 2.6	1.3	0.7– 2.7	1.7	0.9– 2.6	1.2	0.2– 1.7	1.8	0.8– 3.5
Ignition loss	10.6	6.8–17.1	9.3	5.3–12.6	9.0	3.9–13.1	10.9	4.4–14.8	8.9	5.6–19.6
SO_3	0.4	0.1– 1.0	0.5	0.2– 1.1	0.2	0.1– 0.4	0.3	0.1– 0.4	0.3	0.1– 1.8
Carbon	1.6	0.1– 5.7	0.9	0.3– 2.1	0.9	0.3– 2.2	1.7	0.3– 4.8	1.2	0.2– 5.6
Kaolinite	53.3	35.6–75.6	52.6	28.7–68.4	50.4	23.5–67.8	55.5	23.4–93.3	42.9	26.9–67.6
Mica	14.8	6.5–25.8	12.6	6.5–27.8	15.7	7.6–24.0	12.1	2.2–28.9	17.7	5.8–30.6
Quartz	25.4	9.0–44.6	30.0	19.0–58.0	29.3	10.4–67.0	26.2	0.1–65.6	33.1	4.7–50.8
Carbonaceous matter	2.2	0.2– 7.7	1.3	0.0– 3.3	1.2	0.0– 3.0	2.6	0.5– 6.1	2.0	0.3– 9.9

Geology of the Clays

Particle size (%)										
<0.1 μ	9	4–12	4	3–6	3	2–7	3	2.5–7	—	—
0.1–2.0	38	9–47	24	17–30	22	9–30	29	13–53	49.8	33.0–83.5
2.0–5.0	19	10–26	13	8–20	13	8–18	14	3–23	22	8–35
5.0–10.0	9	1–15	10	6–28	11	5–18	9	0–19	12	2–21
10.0–25.0	10	5–17	12	8–17	14	4–20	10	1–16	10	0–31
Critical moisture content (%)	11.0	9.0–15.1	10.1	9.1–11.5	9.6	8.5–11.4	10.0	8.5–12.1	11.5	9.1–15.3
Modulus of rupture (unfired) (MN m^{-2})	2.16	1.07–3.72	2.10	1.17–3.31	1.83	1.31–3.45	2.19	1.24–4.07	2.96	1.52–5.79
Refractoriness (°C)	1647	1530–1710	1630	1460–1680	1662	1595–1700	1660	1490–>1770	1595	1410–1720
Linear firing-shrinkage (%)	4.6	2.0–6.0	2.8	0.0–4.0	2.8	0.0–5.0	3.2	0.5–5.0	4.2	0.9–8.5

Deflocculation

The theory of deflocculation is discussed later, but some practical aspects should be mentioned at this stage. The most commonly used deflocculants for fireclays when the latter are used for casting is a 1:1 mixture by weight of sodium carbonate and sodium silicate. This appears to work satisfactorily provided the exchangeable cations associated with the clay are mainly calcium, i.e. a Ca-clay. However, the author has examined a local fireclay seam that contained Ca-clay at one end but gradually merged into H-clay as one worked towards the other end. The H-clay could not be deflocculated with the normal 1:1 deflocculant, and even though deflocculation could be achieved with a higher ratio of silicate or with NaOH, it was not found possible to produce a satisfactory casting slip with the H-clay.

In the determination of size distribution, the deflocculant used is almost always 'Calgon' (sodium hexametaphosphate), or a 1:1 mixture of 'Calgon' and NaOH. These deflocculants are very effective in the short term but may not be stable enough to be used in casting slips.

Refractoriness

The refractoriness of a clay is carried out by the standard *pyrometric cone test* and the refractoriness expressed as a cone number or (if the test has been properly carried out under controlled conditions) a temperature in degrees Centigrade. A clay suitable for the manufacture of firebricks should have a refractoriness of cone 30, corresponding to a temperature of 1670°C.

As already pointed out, the available data do not include the clays of highest refractoriness, but from the values of Table 14 it is noteworthy that the Scottish fireclays cover the largest range of refractoriness and include the highest value found (over 1770°C). The Staffordshire and Shropshire fireclays, used for making glazed tiles, have the lowest mean refractoriness of the five groups. It is also interesting that the most refractory clay just referred to corresponds to the highest alumina content in the series, viz. 37·8%.

Firing shrinkage

Clays in general shrink when fired because the pore spaces gradually become filled with molten material which on cooling solidifies to form a glass. If the firing is sufficiently prolonged, all the pore space is filled and the body is then said to be *vitreous* and has no porosity.

Other non-clay materials can affect the shrinkage very markedly, however. Prefired material ('grog') is added to fireclay for the purpose of reducing the shrinkage. Since a considerable expansion occurs when quartz is converted to tridymite or cristobalite, a high proportion of quartz in a clay will reduce the firing contraction. High alkali contents (e.g. in the form of mica) increase the firing contraction because the alkali oxides are strong fluxes and therefore produce a high proportion of liquid at the firing temperature.

Since the reactions that occur during firing never go to completion, the size distribution of the clay must also influence the firing contraction. For example, if the particles of mica are fairly coarse, not all the alkali may be released during the firing since the ions have to diffuse through a considerable thickness; for fine particles, the fluxing action of the alkali will be faster because the diffusing ion has a shorter distance to travel.

Bloating

Certain clays when heated to a high temperature not only start to vitrify but expand very markedly. Such clays, whilst not suitable as load-bearing refractories, are nevertheless invaluable as ladle bricks for the steel industry, because the expansion seals the joints between adjacent bricks.

In order to bloat at high temperatures, a clay must in general contain substances that evolve gas, notably sulphates, sulphides, carbonates or fluorides. Equally important, however, it must be fusible enough to form an impermeable layer on the outside of the brick to prevent escape of gas. Thus, the viscosity of the liquid formed at high temperatures is probably critical; if too low, the gases may be able to force their way out and no bloating occurs; if too high, pressure may build up without causing bloating. It has also been stated that carbonaceous matter can produce low-temperature bloating if vitrification occurs before complete combustion of carbon.

Attempts have been made, from time to time, to produce bloating artificially in a refractory clay by the addition of substances such as calcium sulphate, but these are not always successful because so many critical factors are involved.

Vitrification

Of the fireclays examined, none of the South Midlands, Yorkshire, Northumberland, Durham and Scottish groups had vitrified at 1200°C, the average porosity after firing being around 20%. By contrast, of the Staffordshire and Shropshire group, 6 samples out of a total of 64 had completely vitrified at 1200°C, which is consistent with their lower average refractoriness.

Plasticity

It is generally considered that the plasticity of fireclays is on average intermediate between that of china clay and ball clay, a conclusion that is borne out by recent tests using the compression plastometer and other methods. The underclays, i.e. those occurring immediately beneath the coal, are more plastic than the shales. The plasticity of fireclays may be enhanced by weathering, which breaks up the hard, compacted grains, making them accessible to water.

Unlike ball clays, fireclays are contaminated with organic matter akin to bituminous coal, which does not contribute to plasticity.

Brick clays

Occurrence in the United Kingdom

Whereas the requirements for a whiteware or refractory clay are fairly stringent, those for a brick-making clay are much less so. A wide range of compositions can therefore be tolerated in brick-making clays and so sedimentary-type clays of various geological ages can be utilised. It is convenient therefore to classify the brick clays according to the geological formations in which they occur (see Table 8).

Devonian deposits. Clay deposits of the Devonian period are used for brick-making in South Devon (near Torquay), Cornwall, and around Cardiff and Newport.

Carboniferous deposits. Non-refractory clays of the Carboniferous period, particularly those of the Coal Measures, are used to a considerable extent in the brick-making industry. Suitable brick clays of this type are worked in the Glasgow district and in the neighbourhoods of Durham and Newcastle-on-Tyne. In North Staffordshire, clays of the Upper and Lower Coal Measures, including the so-called Etruria Marls, are used for the manufacture of bricks and roofing-tiles. The sequence of seams in this area, after removal of overburden, consists first of sandstones and ironstones, followed immediately below by shales and grey sandstones, underneath which lies the Etruria Marl, followed by the Black Band group of clays. The quality of these clays varies considerably from place to place but the deposits are mostly hard and require weathering. Other Coal Measure clays utilised for brick-making occur in Leeds, Halifax, Huddersfield and Wakefield (Yorkshire), Mansfield (Nottinghamshire), Ashby-de-la-Zouch and Measham (Leicestershire), Nuneaton (Warwickshire), Tipton, Bilston, Willenhall and Walsall (West Midlands), Bristol (Avon), Cardiff (Glamorganshire) and Ruabon (Clwyd).

The Keuper Marls. This important group of clays of the Triassic period extends across the country from Sidmouth (Devon) to the mouth of the Tees and is utilised for brick-making near Birmingham, Mapperley (Nottinghamshire) and Leicester. Their use is often limited because of contamination with rock-salt and gypsum ($CaSO_4.2H_2O$).

The Oxford clay. This is a Jurassic deposit, rich in organic matter, that is worked at Bletchley (Buckinghamshire), Kempston, Hardwick, Leighton Buzzard and Stewartby (Bedfordshire), and at Peterborough (Cambridgeshire). Some 40% of the total brick production of the UK is made from this clay.

The Hastings Beds. These are clays of Lower Cretaceous origin, associated with alternate deposits of sand. They are used for brick-making at Hastings and Bexhill.

The Weald clay. Also of the Lower Cretaceous period, the Weald clay occurs in Sussex, the Isle of Purbeck (Dorset) and in parts of the Isle of Wight, and is worked in all these areas.

The Gault clay. Forming part of the Upper Cretaceous system, the Gault forms a thin, straggling line across Dorset and Wiltshire, running from Shaftesbury (Dorset) to north of Abbotsbury (Dorset); it is located again at Westbury, can be traced as far as Devizes (Wiltshire) and outcrops again in east Hampshire, Berkshire, Oxfordshire, Buckinghamshire, Bedfordshire and Cambridgeshire. Part of it also branches across Surrey, Sussex and Kent. The Gault is a bluish-grey to bluish-black marine clay, and is utilised for the manufacture of bricks, drainpipes and tiles near Eastbourne and for brick-making at Shaftesbury and Devizes.

Eocene deposits. Clays of Eocene age occur in two principal areas, the Hampshire Basin and the London Basin. These deposits are used for brick-making in various localities, and include the clays of the Reading Beds, the London clay, the Bracklesham Beds (Isle of Wight) and the Barton clay (Hampshire).

Oligocene deposits. These deposits occur only in the Hampshire Basin and in Devonshire. Among those used are part of the Headon Beds near Bournemouth and the Hampstead Beds (Isle of Wight).

Pliocene deposits. Deposits of this system are worked in a few places in East Anglia (Norfolk and Suffolk). Many are unsuitable for brick-making because they are too calcareous and fusible.

Pleistocene and Holocene deposits. These deposits include the *alluvium*, a deposit found in certain river valleys, notably the Humber estuary and Lincolnshire coast, the Vale of Pickering and the Vale of York, the Thames and Medway valleys and around Bridgwater. The *brick-earth* is a loamy deposit found in river valleys, and on the slopes and terraces. They occur in the Thames valley, north Kent and in the estuaries of the Forth, Tay and Clyde. Many former brick-earth deposits have been almost completely worked out.

Boulder clay is of glacial origin, of the Holocene or Recent period, and, as the name implies, is associated with boulders and small stones. This clay is scattered over East Anglia, Leicestershire, Staffordshire, Shropshire, Warwickshire, Wales, the North of England and the South of Scotland. Brick-works utilising Boulder clay are situated at Buckley (Clwyd), near Manchester, at Peel (Isle of Man) and at Sudbury (Suffolk).

Other brick clay deposits

In Canada, the less refractory clays of the Saskatchewan Whitemud Formation have been used for the manufacture of facing bricks. Facing bricks are also made from Cretaceous clays found at Kergwenan, Manitoba, and from clays of similar age occurring in Nova Scotia.

In South Africa, high-quality brick clays are found in the Transvaal, within a fifty-mile radius of Johannesburg; in the Western Cape near the Cape Peninsula; in Natal, around Durban and Pietermaritzburg; at Bloemfontein, Orange Free State; and at Grahamstown, in the Eastern Cape.

Composition of brick clays

The principal minerals in brick clays are disordered kaolinite, chlorite and illite, with quartz and organic matter. Brick clays have been examined that appear to contain little or no kaolinite but may depend for their clay-like properties on chlorite or illite. In addition, many oxides, carbonates, sulphides and other impurities may be present. Because the composition of the minerals present, particularly chlorite, is uncertain, it is difficult to calculate a 'rational analysis' for brick clays, as may be done for ball clays.

From the chemical compositions of bricks made from clays of various ages (Table 15) it can be seen that although the range of compositions is wide, most clays contain appreciable proportions of iron oxide and calcium oxide. The latter is reported as CaO in analyses but in the raw clay it is probably present as $CaCO_3$. Soluble salts, chiefly calcium sulphate, sometimes occur in these clays and can cause efflorescence in bricks made from them. Some of the more important properties of a selection of brick clays are given in Table 16.

Particle size distribution

As the figures of Table 16 show, all the clays examined contain an appreciable amount of material less than 2 μm in diameter, the latter range usually being considered as the 'clay fraction'. Thus, the clay minerals are intrinsically fine-grained and the coarse material is contributed by the impurities. The value quoted for the percentage less than 2 μm is the average of between 50 and 100 clays of each formation.

Deflocculation

The response of a brick clay to deflocculants is of little technological interest, since such clays are not used in casting; deflocculants are used, however, in determining size distribution by sedimentation methods and it is of interest that, for the Etruria Marls, a mixture of five parts by weight of 'Calgon' to one part of NaOH was a more effective deflocculant than sodium oxalate. All other clays in this survey were adequately deflocculated with the same combination of reagents. This behaviour is probably connected with the exchangeable cations associated with the clays, as explained later.

Working moisture content

The working moisture content is an arbitrary measurement, defined, in this instance, as the moisture content required to produce the correct consistency for extrusion. As explained later, the moisture content corresponding to a given consistency is governed by a large number of factors, such as mineralogical composition, size distribution in the 0–2 μm range, nature of exchangeable cations, etc., and it is therefore difficult to predict. However, it is the working moisture content that largely determines the amount of wet-to-dry shrinkage that occurs in any plastic forming process.

Table 15

Analyses of Bricks made from Clays of Various Geological Deposits (%)
(From *Clay Building Bricks of the United Kingdom*, by Bonnell and Butterworth, HMSO, 1950)

	Alluvial	London stock	Glacial	Oligocene	London clay	Gault	Weald	Oxford clay (Fletton)	Middle Lias	Keuper Marl (Upper)	Keuper Marl (Middle)	Keuper Marl (Lower)	Permian	Etruria Marl	Coal measure	Coal measure	Coal measure	Devonian
SiO_2	64.7	68.7	62.5	77.8	64.4	47.2	68.4	56.2	57.8	58.3	46.2	66.0	60.1	62.7	54.9	61.7	61.9	59.6
Al_2O_3	12.7	11.0	18.6	15.8	15.8	19.4	17.2	20.9	23.2	15.3	13.7	13.9	16.7	23.1	34.9	21.6	24.0	19.9
Fe_2O_3	8.3	7.0	6.6	0.8	7.9	6.1	6.3	6.0	9.3	6.0	6.0	6.8	5.8	8.4	3.4	8.0	8.7	11.4
TiO_2	1.6	0.7	0.9	2.2	1.6	0.8	1.3	0.5	1.2	0.7	0.8	0.6	0.5	1.2	0.6	1.2	1.1	1.2
CaO	7.9	8.1	4.1	0.3	1.1	19.2	1.9	8.1	1.0	6.2	11.4	3.5	6.5	0.9	2.1	0.6	0.6	0.2
MgO	1.9	0.8	3.4	0.4	2.4	1.9	1.2	1.7	2.5	7.3	12.8	2.9	4.2	1.2	0.7	1.0	1.1	1.2
Na_2O	0.4	0.8	0.5	0.5	0.5	0.6	0.5	0.5	0.9	0.7	0.3	0.6	1.2	0.4	0.1	1.2	0.2	1.0
K_2O	1.5	2.0	2.9	1.8	3.2	3.1	2.2	3.6	2.9	4.7	3.3	4.1	3.3	2.6	2.6	3.1	1.6	4.2
SO_3	1.4	0.6	0.4	0.5	2.3	1.4	0.7	1.9	0.3	0.5	5.6	1.0	0.9	0.7	0.1	1.5	—	0.1
Loss	0.3	nil	0.3	0.3	1.1	0.4	0.3	0.6	1.0	0.4	0.5	0.2	0.3	nil	0.6	0.1	0.5	0.5

Table 16
Physical Properties of some Brick Clays

Clay	Particle size (% smaller than 2 μm diameter)		Working moisture content (% dry basis)		Linear firing shrinkage (%)		Fired colour for different temperatures	
	Range	Average	Range	Average	Temperature (°C)	Range	Temperature (°C)	Colour obtained
Coal Measure shale (outcrop)	16–78	34	15·1–31·2	19·5	1050° 1180°	0·2–7·7 2·0–9·7	1050° 1180°	Light salmon, red pink light cream Green-grey, dark brown, chocolate brown, greenish yellow
Coal Measure shale (from coal seams)	14–43	26	15·0–25·1	17·8	1000° 1180°	0·4–5·1 0·7–7·7	1000° 1180°	Salmon pink, pale pink, buff, white Green-grey, light grey, cream
Etruria Marls	24–74	47	15·2–30·2	23·6	1180°	0·6–12·1	1000° 1180°	Light buff, pink buff, light red, light brown, light chocolate brown, pink, cream Dull brown, red brown, purple brown, medium red, stone, greyish buff
Weald clay	15–85	47	23·8–42·2	30·9	900° 1100°	0–3·4 2·5–8·6	850° 1200°	Salmon pink, light red-brown, buff Dark red-brown
Boulder clays	30–60	47	17·1–39·6	28·8	900° 1070°	0·1–4·3 2·5–9·4	880° 1070°	Salmon pink, light brown Light chocolate brown, light brown

Firing shrinkage

To obtain a valid comparison of the firing shrinkages of different clays, a constant firing temperature would have to be employed for all specimens; this is impossible in practice, because the fusibility varies so widely that some clays would vitrify at a temperature too low to cause appreciable shrinkage in others. For this reason, the firing shrinkages quoted in Table 16 were measured after firing at temperatures suited to the particular clay and thus show some variation. In particular, the Boulder clays, being calcareous, were more readily fusible than most and showed appreciable shrinkage at as low as 900°C. A further difficulty is that some clays bloat at around 1200°C, introducing a serious source of error into the measurements; where bloating is detected, of course, the specimens in question are rejected.

Since the firing shrinkage is caused by reactions associated with the clay minerals, it can be correlated approximately with the percentage of material below 2 μm.

Fired colour

The fired colour of brick clays is due chiefly to oxides of iron, but the colour is modified appreciably by the other constituents present. Ferric oxide, Fe_2O_3, the normal product of firing, produces various shades of red and yellow; firing under reducing conditions may form ferrous compounds or Fe_3O_4, resulting in blue or black colours. At relatively high temperatures (around 1300°C) Fe_2O_3 may dissociate to Fe_3O_4 and oxygen, giving a dark-brown colour unless reoxidised during cooling. Not only the proportion of iron oxide, but the grain size, intimacy of mixing, the presence of other constituents and the firing temperature can all influence the colour.

In discussing the effect of other constituents on fired colour, it is convenient to consider three groups, based on composition:

(a) High iron oxide (5–9%), low Al_2O_3 (10–22%), CaO negligible. In this range of composition, all shades of red are obtained, the shade darkening as the firing temperature is raised.

(b) Fe_2O_3 lower (1–3%), Al_2O_3 high (about 25%). With this composition, there is a pronounced modifying effect by the alumina, giving a buff colour when cooled. The colour is pink at 500–730°C, faint pink at 825°C and almost white at 1000°C. At 1100°C a faint yellow-green is produced, reaching maximum development at about 1220°C and lightening again at 1320°C, becoming light brown on slow cooling. The buff colour is said to be due to formation of a solid solution of Fe_2O_3 in mullite and the pinks to free Fe_2O_3. For very low iron oxide contents (less than about 1%) the effects are similar to those above, giving pink or white at low temperatures and cream at high temperatures.

(c) Calcareous clays (high CaO content). The natural red or brown colour of ferric oxide is bleached strongly by CaO. If the $CaO:Fe_2O_3$ content is greater than 2:1, as in the Gault clays, cream and light yellow colours are formed. A ratio of 1·5 to 1·8 forms yellows under slightly reducing conditions, but under fully oxidising conditions pinks or reds can be formed. The bleaching action of CaO is inhibited if oxides of sulphur (derived from the kiln) are present, since the CaO is then converted to $CaSO_4$.

The darkening effect of increased firing temperature is well illustrated by the data of Table 16. The light colours obtained with the Coal Measure clays are probably a consequence of the relatively high alumina content (cf. Table 15) which also has some bleaching action. The moderately high CaO content of the Boulder clays (about 4%) is apparently sufficient to produce light shades of colour, but under oxidising conditions is not high enough to ensure a yellow colour.

Vitrification

Since the oxides of sodium, potassium, calcium and magnesium act as fluxes and promote vitrification, brick clays containing a high proportion of $CaCO_3$ are readily fusible and have therefore to be fired at a relatively low temperature. The Gault clay and the Keuper Marl are both noted for their lime content and fusibility.

Extraction of brick clays

Brick clays are mostly extracted from open workings, but if they outcrop on the side of a hill, with an extensive overburden, the most convenient method of extraction is to drive a tunnel into the side of the hill at the level of the seam (drift or tunnel mining). If the seam is not more than 20 ft (6 m) below the surface, the overburden is stripped off by means of skimmers, power shovels, bulldozers or shoveldozers, and the underlying clay is removed with a dragline scraper or skimmer (Fig. 33). Sometimes shallow pits are dug out and the clay is stripped from the sides of the pit with a bucket-type mechanical excavator (Fig. 34).

Fig. 33. Removing overburden.

Fig. 34. A bucket excavator (photograph by courtesy of the London Brick Company Ltd).

Stoneware clays

These are siliceous, readily fusible plastic clays that are not white-burning. Some non-refractory clays of the Coal Measures are used as stoneware clays.

Pipe-clays

These are similar in nature to brick clays, being plastic, fusible, red-burning clays, deposits of which are found in various geological formations.

Bentonite

The name *bentonite* is a general term given to a swelling-type montmorillonite clay, containing sodium as the predominant exchangeable cation. It occurs in the United States as a Cretaceous deposit, formed by the alteration of volcanic ash, in Wyoming, South Dakota, Nevada and Mississippi. The name 'non-swelling bentonite' has been given to a calcium montmorillonite found at Panther Creek, north-eastern Mississippi. In Canada, Cretaceous deposits of bentonite occur in Manitoba, Saskatchewan and Alberta; and Eocene deposits in British Columbia.

Fuller's earth

Although bentonite is not found in the United Kingdom, a calcium montmorillonite known as 'fuller's earth' is mined east of Redhill, Surrey; this is a Cretaceous deposit of the Lower Greensand. Another deposit of the Lower Greensand is worked near Woburn, Bedfordshire. Good commercial fuller's earth also occurs as a Middle Jurassic deposit near Bath, Avon.

Fuller's earth is so called because it was originally used for 'fulling', i.e. removing the fatty substances from wool. Because of its large surface area, high cation exchange capacity and adsorptive capacity, this clay now finds use in water and effluent treatment, in refining and clarification of sugar solutions, syrups and wines, in paints, glazes, foundry moulding sands and as a plasticiser in ceramic bodies. Fuller's earth is obtained by both opencast and underground mining methods.

REFERENCE

EXLEY, C., *Clay Min.*, **11** (1976) 51.

READING LIST

D. V. AGER, *Introducing Geology*, Faber, 1961.
F. H. CLEWS, *Heavy Clay Technology*, Academic Press, 1968.
A. HOLMES, *Principles of Physical Geology*, Nelson, 1965.
P. S. KEELING, *The Geology and Mineralogy of Brick Clays*, Brick Development Association Monograph, 1963.
I. F. KIRKALDY, *Minerals and Rocks*, Blandford Press, 1968.

Chapter 5

Properties of Clay–Water Systems

THE IMPORTANCE OF CLAY–WATER SYSTEMS

One of the characteristics of a clay is that when mixed with water it forms a coherent mass that is capable of being moulded to any desired shape, i.e. it is *plastic*. It will be noted that many other substances, even when finely powdered, do not possess this property and are *non-plastic*, e.g. sand, flint, chalk, etc. If sufficient water is added to clay, a suspension (or slip) is formed; on removal of some of the water, a putty-like mass or paste is formed. The unique characteristics of clay in contact with water are due to its *colloidal properties*. In order to understand the behaviour of clays, it is therefore necessary to study colloidal systems in general.

COLLOIDS

It is a common experience that when clay is shaken and dispersed in water, the resulting suspension remains cloudy on standing and frequently may not clear for days or even weeks. This is because a *colloidal solution* of clay in water has been formed, the particles of clay being so small that they settle extremely slowly, if at all. Many other examples of colloidal solutions can be given; e.g. on passing hydrogen sulphide into an oxidising solution, sulphur may be precipitated in colloidal form. Again, the familiar Prussian blue, obtained by adding a solution of potassium ferrocyanide to one of ferric chloride, is also a colloidal solution.

The examples quoted are of the solid–liquid type, but in fact there exist colloidal systems of solid in gas (e.g. smoke), liquid in gas (e.g. aerosols), liquid in liquid (e.g. emulsions) and solid in solid (e.g. certain coloured glasses). The particulate substance is known as the *disperse phase*, the suspending medium as the *continuous phase*. We therefore define a colloid in a very general way, namely, a phase dispersed to such a degree that its surface properties are predominant. Although there is no sharp distinction between ordinary particulate matter and colloids, it is generally accepted that colloidal properties become pronounced for particles smaller than about 0·2 μm in diameter. By way of illustration, consider a cube of solid material, of side 1 mm and density 2·5 g cm^{-3}. The specific surface area of this cube (i.e. surface area per unit mass) is equal to 24 cm^2 g^{-1}. The same material, in the form of cubes of 0·2 μm side, has a surface area of 120,000 cm^2 g^{-1}. The enormous surface area of the colloidal-sized unit cube is the main reason for its characteristic properties. For the sake of comparison, the size of atoms and molecules is of the order of 0·0005 μm or 0·5 nm.

Another important aspect of colloidal suspensions is their ability to remain in suspension indefinitely. For a spherical particle suspended in a liquid, Stokes' Law states that the velocity of sedimentation, v, is given by the equation:

$$v = 2r^2(d_1 - d_2)g/9\eta$$

where r = radius of particle, d_1 = density of solid, d_2 = density of liquid, g = acceleration due to gravity and η = viscosity of liquid. However, small particles suspended in a liquid exert an outward pressure like gas molecules, resulting in a *diffusion* effect which opposes sedimentation. According to Fick's Law of steady-state diffusion, the amount dw of substance diffusing across unit area in time dt is given by the equation:

$$dw = -D\frac{dc}{dx}dt$$

where D = diffusion coefficient and dc/dx = concentration gradient. For a constant temperature and viscosity, the constant D is proportional to $1/r$, where r is again the radius of the particle diffusing. Comparing the equation for sedimentation with that for diffusion, it is clear that the former, being a square law, rapidly becomes very small for particles of colloidal size, whilst the latter increases rapidly as r decreases. Clearly, there is a limiting size for r where diffusion is sufficiently rapid to prevent any great degree of sedimentation; this becomes manifestly so when r is around 0·2 μm. Of course, the discussion has been oversimplified because it has been assumed that all particles are equal-sized spheres, acting as individual units; the possibility of interaction between the particles will be considered later.

Preparation of colloids

Most substances are capable of existing in a colloidal form. In general, the preparation of colloidal suspensions involves either the breakdown or dispersion of large particles into smaller ones, or the build-up of colloidal-sized particles from atoms, ions or molecules.

Dispersion methods

The substance may be ground with the dispersion medium in a mill; or an electric arc may be struck between electrodes made from the same material, if it is electrically conducting. If the substance is already in powder form, merely bringing it into contact with the medium may suffice, particularly if a chemical dispersant is added.

Condensation methods

Where conditions inhibit the growth of large crystals, colloidal suspensions may be prepared by the interaction of two solutions, e.g. by passing H_2S into a solution of arsenious oxide in water, when under suitable conditions colloidal arsenious sulphide, As_2S_3, is formed. Many sulphides and hydroxides may be prepared by chemical precipitation in colloidal form. Where a sulphide or hydroxide has to be filtered, as in chemical analysis, the formation of colloidal suspensions must of course be avoided, since colloidal particles readily pass through even the finest filter papers. Certain animal membranes, however, are capable of retaining particles of colloidal size; filter paper treated with collodion, and likewise porous ceramic vessels, may also be used for filtering colloids.

General properties of colloids

Colloids have been classified as being *lyophilic* (attracted by the dispersion medium) and *lyophobic* (repelled by the dispersion medium). Lyophilic colloids are very stable, absorb the medium and have a high viscosity. Moreover, they are *reversible*, in that they can be evaporated to dryness and then redispersed by merely adding fresh dispersion medium. Examples of lyophilic colloids are common glue, starch and gelatin. Lyophobic colloids, by contrast, are less stable, do not absorb the dispersion medium and are not readily reversible. Examples of lyophobic colloids are colloidal gold, Prussian blue and arsenious sulphide. The distinction between lyophilic and lyophobic colloids is not a sharp one and some substances are intermediate, having some properties of each class. Clays on the whole are lyophobic but some, like the montmorillonites, possess some lyophilic properties. (When water is the dispersion medium, the terms 'hydrophilic' and 'hydrophobic' are sometimes used.)

Colloidal suspensions exhibit the so-called Tyndall effect. If a beam of light, viewed at right angles to its direction, is passed through a colloidal suspension, the light is scattered by the particles, rendering the path of the beam visible. If the beam of light is examined with a microscope, also set at

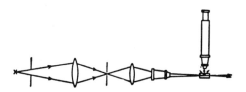

Fig. 35. The ultramicroscope (from Glasstone's *Textbook of Physical Chemistry*, copyright 1946, D. Van Nostrand Co. Inc., New York).

right angles to the beam, the colloidal particles appear as bright points of light, though they are too small to be resolved, i.e. their shape cannot be discerned. This is the principle of the ultramicroscope (Fig. 35).

When examined in this way, the particles are seen to be in rapid, random motion, known as *Brownian motion*. This is caused by molecules of the dispersion medium, which have a thermal motion dependent on the temperature, striking the colloidal particles, causing them to move in a similar manner. Clearly, since the velocity of a particle, for a given kinetic energy, will be inversely proportional to its mass, only very small particles will acquire sufficient velocity for their motion to be detected.

Electrokinetic properties of colloids

If two platinum electrodes, connected to a source of high potential (of the order of 200 V), are immersed in a colloidal suspension, a small current passes and the colloidal particles migrate towards one or other of the electrodes (Fig. 36). This phenomenon is known as *electrophoresis*, and shows that all colloidal particles carry an electric charge. Since the suspension as a whole is electrically neutral, there must clearly be an equal and opposite charge somewhere in the suspension; this equal and opposite charge is provided by various ions (positive or negative), the presence of which can be detected chemically at the opposite electrode. A specific example may make this clearer. If a suspension of clay in water is treated in the above manner, the particles of clay migrate to the anode, showing that they are negatively charged. At the opposite electrode, the cathode, appear the *balancing ions*, or *counter-ions*, which of course are positively charged. With a natural clay, these ions will be a mixture of H, Ca, Mg, Na and K in various proportions, the metallic ions reacting with the water to form hydroxides, the hydrogen appearing as the gas, H_2. Thus, the process

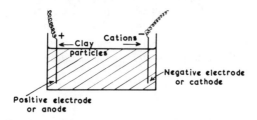

Fig. 36. Electrophoresis.

is somewhat similar to electrolysis. The counter-ions constitute the exchangeable ions referred to in Chapter 3.

The way in which an electric charge develops on colloidal particles is roughly as follows. Where a colloid is formed by precipitation, e.g. the formation of colloidal AgCl by reaction of silver nitrate with a soluble chloride, the AgCl lattice absorbs either silver ions or chloride ions, whichever is in excess. Thus, for silver nitrate and excess sodium chloride, the reaction is:

$$AgNO_3 + NaCl \rightarrow [AgCl]Cl^- + Na^+ + NaNO_3$$
(excess)

In this instance, the silver chloride particles acquire a negative charge by adsorption of chloride, sodium ions acting as counter-ions. If silver nitrate is in excess, the precipitated silver chloride acquires a positive charge by adsorption of silver, with nitrate ions as counter-ions. Note that the 'lattice ion' is firmly bound to the silver chloride, whilst the counter-ion, which cannot enter the lattice because it is of unsuitable size, remains as an exchangeable ion.

With the clay minerals, the colloidal suspensions of which are formed by dispersion and not by precipitation, the balancing ions are already present as exchangeable ions, which either satisfy free valencies at the crystal edges or balance inherent lattice charges due to isomorphous substitution. Therefore, the charge on the clay particles can develop only by diffusion of the counter-ion away from the solid surface when the clay is placed in water:

$$[\text{Na clay}] \xrightarrow{H_2O} [\text{clay}]^- + \text{Na}^+$$

The extent to which diffusion occurs depends on the nature of the cation, on its hydration and on the concentration. The significance of the ion distribution which results from this diffusion is discussed shortly.

Whilst it is likely that most silicates acquire electrical charges in the same manner as do clays, oxides in general derive their charge by direct ionisation. Thus, silica gel (hydrated silica, $SiO_2 \cdot xH_2O$) has Si—OH groups at the solid surface, which ionise normally to produce a negative colloid:

$$[\text{Si—OH}] \rightarrow [\text{Si—O}]^- + H^+$$

Amphoteric oxides, such as alumina, ionise in alkaline environments to give a negative colloid and in acid environments to give a positive colloid:

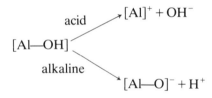

For such substances there will clearly be some intermediate pH where neither mechanism predominates, so that the charge is zero; the pH at which this occurs is known as the *isoelectric point* (IEP). Estimates of the IEP for $\alpha\text{-}Al_2O_3$ vary but the commonest reported value is about pH 5·5. It should be noted that the surfaces of all oxides in contact with water become hydrated and thus the surface of alumina is effectively Al—OH, as shown. Basic oxides in general ionise to produce a positive colloid, but it is possible in some instances to reverse the charge at very high pH values, i.e. the IEP occurs at a very high pH value.

The stability of colloids

Provided the particles of a colloidal suspension remain as discrete individual units, Brownian motion and diffusion operate to counteract the sedimentation forces. If, however, the particles are caused to gather together into flocs or aggregates, the latter behave as single, large particles, sedimentation occurs and the system is said to be *flocculated*. Thus, a colloidal system is unstable if the particles composing it have a tendency to aggregate. Owing to their thermal motion, particles frequently come into collision with each other; whether such collisions result in the particles joining together depends on the electrical forces between them.

First, there are short-range forces of attraction, which will invariably operate to cause aggregation if any two particles approach closely enough. These short-range attraction forces operate with all solid surfaces, though they may be considerably reduced, particularly with lyophilic colloids, by *solvation*, i.e. the affinity of the solid surface for the solvent.

Second, there are the electrostatic forces of repulsion associated with the charges on the solid surface. The amount of work done in bringing two charged particles together depends on their electric potentials; the higher the latter, the greater the amount of work required and the less likelihood there is of aggregation. The important factor is therefore not the charge itself but the potential due to the entire assembly of the charged surface and its counter-ions; this is shown diagrammatically in Fig. 37. The vertical line on the extreme left of the diagram represents the boundary of a solid (clay) surface, in contact with water with an associated charge (negative in this instance). The positive counter-ions are distributed as shown, and the electrical potential is plotted on the same diagram against distance from the solid surface. On the currently accepted *Stern theory*, a fraction of the positive counter-ions are strongly adsorbed and form a monolayer close to the surface, known as the *Stern layer*, the boundary of which is indicated by the vertical dotted line. As can be seen from the curve, there is a sharp fall of potential within this layer, due to the high concentration of counter-ions, from the value ψ_0 at the surface to ψ_d at the boundary of the Stern layer. ψ_0 is sometimes called the Nernst potential, and ψ_d is referred to as the *Stern potential*. It is the latter that is important in colloidal systems, since unlike the Nernst potential it is very sensitive to changes in the nature and concentration of the counter-ions. Those counter-ions outside the Stern layer form an 'atmosphere' adjacent to the solid surface, their concentration falling off exponentially with distance from the surface (Fig. 37).

By virtue of the electrical charge on the surface, the latter adsorbs water molecules, which are thus held by the surface as an adherent film, called the *lyosphere* for liquids in general, or *hydrosphere* in the case of water. This adsorption of the solvent can occur with *polar liquids*, i.e. those in

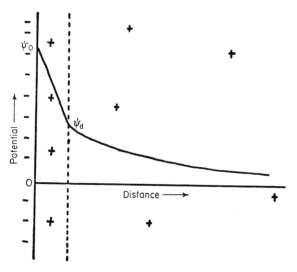

Fig. 37. Distribution of ions on a charged particle.

which the positive and negative centres of the molecules do not coincide, resulting in a partial separation of charges. The molecule then behaves rather like a small magnet, one end (in this instance the positive end) being attracted to the solid surface, the other pointing away from it. Polar molecules (or dipoles) are formed by the combination of elements of slightly differing electronegativities, so that the bonding, although essentially covalent, has some degree of ionic character. Thus, water, alcohols, acetone, etc., are polar liquids, whereas the paraffins are non-polar.

The thickness of the adsorbed water layer has long been a matter of dispute, but it now seems that tightly bound molecules do not extend much beyond the Stern layer. When the particle migrates, the water layer, together with the ions in the Stern layer, moves with it. Therefore, the electrical interaction between adjacent particles depends on the potential at the boundary of the immobile water layer, usually known as the zeta- (or electrokinetic) potential (ζ). In practice, therefore, it is the zeta-potential which governs the stability of colloidal systems; if of sufficient magnitude, it can effectively prevent aggregation of two or more adjacent particles. At the same time, it should be realised that the zeta-potential is in turn dependent on the Stern potential and is just marginally smaller than the latter.

According to the theory developed independently by Deryagin and

Landau (1941) and Verwey and Overbeek (1948), the net interaction between two flat colloidal particles can be written as:

$$E = E_R + E_A$$

where E = total energy of interaction, E_A = energy of attraction and E_R = energy of repulsion. Further calculation shows that:

$$E = -A\exp(-kx) + B/x$$

where x = separation distance, A = a factor dependent largely on the Stern potential, k = a constant related to electrolyte concentration, B = a constant related only to the nature of the solid and the suspension medium. It will be clear that the first term, E_R, represents the electrical *repulsion* forces (with a negative sign) whilst the second term E_A denotes the short-range van der Waals-type forces of attraction. It can furthermore be shown that below a critical Stern potential (i.e. low values of A) the summation of these two terms is positive, so that a net attraction exists at all distances of separation; above this value there exists a range of separation distances x over which there is a net repulsion. Thus, provided the Stern potential is sufficiently high (in practice some 50 mV or over) this region of repulsion acts as a potential barrier, effectively preventing too close an approach of any two particles and so preventing them forming aggregates.

The zeta-potential varies with the nature and concentration of the ions adjacent to the surface, particularly the counter-ions. For low concentrations, the alkali ions and ammonium produce high zeta-potentials as counter-ions, the order of magnitude being $Li > Na > K > NH_4$. Divalent and polyvalent ions produce low zeta-potentials, the order being $Mg > H > Ca$, although some clays do occasionally give somewhat different orders within either series. High concentrations of any ion cause the Stern layer to become 'crowded' and so reduce the Stern and zeta-potentials.

No completely satisfactory explanation for the behaviour of different cations has yet been put forward. It is generally accepted that ions in solution are hydrated, i.e. they carry with them an envelope of adsorbed water molecules. On this basis, it has been suggested that the number of ions that can be accommodated in the Stern layer is governed by the size of the *hydrated* ion; the larger the hydrated ion, the fewer that can be accommodated as a tightly adsorbed monomolecular layer and hence the higher the potential at the Stern boundary, since the drop in potential is not then so abrupt (Fig. 37). This explanation accounts for differences between the monovalent group, and in general for differences between ions of the same valency, but breaks down altogether when ions of different valency are compared. This will be appreciated from the values of Table 17, where average values for degree of hydration, and hydrated and non-hydrated radii, for various cations are compared.

The alkali series, Li, Na and K, have hydrated radii that decrease in the order quoted, in conformity with the respective zeta-potentials. Mg, with a much larger hydrated radius than any of the alkalis, is always associated with a low zeta-potential. Even when the valency is taken into consideration, the extra positive charge carried by the Mg ion would seem scarcely

Table 17
The Effect of Hydration on Ionic Radius

Ion	Normal radius (Å)	Hydration (mol H_2O)	Hydrated radius (Å)
Li	0.78	14	7.3
Na	0.98	10	5.6
K	1.33	6	3.8
NH_4	1.43	3	—
Rb	1.49	0.5	3.6
Cs	1.65	0.2	3.6
Mg	0.78	22	10.8
Ca	1.06	20	9.6
Ba	1.43	19	8.8
Al	0.57	57	—

sufficient to compensate for the relatively small number of hydrated Mg ions that could be accommodated in the Stern layer. It must be admitted, however, that the values quoted for hydration are usually those obtained for ions in solution, but there is some evidence that ions in close proximity to a charged surface may hydrate to a different extent from those in solution. Some authorities have ignored hydration, and have suggested that the tendency for an ion to remain in the Stern layer depends on its electropositivity. Thus, the negatively charged clay particle is regarded as the anion of a weak acid, the associated cations forming the corresponding 'salt'. The more electropositive the cation, the more it will tend to ionise and diffuse away from the solid surface:

$$Na[clay] \rightarrow [clay]^- + Na^+$$

This hypothesis accounts for the difference between, say, Na and Mg, but

Properties of Clay–Water Systems

reverses the order for the alkali metals, since Na is more electropositive than Li, yet the latter produces a higher zeta-potential. Yet another suggestion is that the dominant factor is the polarisability of the associated cation, a quantity dependent on charge/ionic radius, but like the electropositivity, to which it is closely related, this does not account satisfactorily for the behaviour of different ions. More information on the hydration of ions attached to surfaces would be helpful in this connection, but in any event it seems likely that several different factors are involved in determining the precise effect of any one ion on zeta-potential.

On the Stern theory, it is clearly possible for the Stern or zeta-potential to be of opposite sign to the Nernst potential, if a sufficient number of counter-ions can be accommodated in the Stern layer. Because of their inherent lattice charge, this is clearly less likely to occur with clays than with other substances, but instances have been reported where clays have acquired a positive zeta-potential by adsorption of tetravalent cations. An alternative explanation is that polyvalent cations are very prone to form basic ions of type M—OH, so that the surface of the clay becomes a hydroxyl surface, the latter being exchangeable for other ions by ionisation:

$$[\text{clay M—OH}] \rightarrow [\text{clay M}]^+ + \text{OH}^-$$

When one considers colloidal suspensions in non-aqueous media, information is relatively scarce. Colloidal particles can undoubtedly acquire zeta-potentials in non-aqueous media, but the mechanism is not fully understood. Some authorities have expressed the view that such charges are frictional in origin, like electrostatic charges. On the other hand, it has been observed that clays can be stabilised in non-aqueous media by adsorption of polar molecules of large charge separation, such as acetic acid, trichloracetic acid, etc. These adsorbed molecules are said to form a polar screen around the particles, preventing aggregation, but it is difficult to understand how this screen operates with two sets of charges that are relatively fixed. It is equally probable that the effective stabilisation of clay in non-aqueous media by polar molecules is due to the increase in solvation energy resulting from more-or-less complete coverage of the clay surface by an organic molecule.

Determination of zeta-potential

Because the zeta-potential, as presently defined, is the potential at the boundary between the 'immobile' water layer and the bulk of the suspending medium, it is the only quantity that can be directly determined experimentally.

Consider a charged particle (Fig. 38) separated from an equal and oppositely charged layer by a distance d, forming an electrical condenser. This simplified model, due to Helmholtz, considers the counter-ions as falling entirely within one plane, which is of course only approximately true. If the charge density on each layer is σ and the dielectric constant of

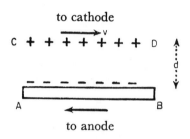

Fig. 38. Migration of a charged particle.

the medium is ϵ, then from electrostatics the capacity of this parallel-plate condenser is

$$\frac{\epsilon\epsilon_0 A}{d}$$

where A = area of the plate. But potential ζ = total charge/capacity

hence $\quad \zeta = \sigma A / \dfrac{\epsilon\epsilon_0 A}{d}$

i.e. $\quad \zeta = d\sigma/\epsilon\epsilon_0 \hfill (1)$

The various methods of determining zeta-potential are based on this simple relationship. They depend on causing a colloidal particle to move under the influence of an electrostatic field and measuring the rate of migration; or by measuring the so-called *streaming potential* which results when the liquid medium is made to flow under pressure through a porous plug of the colloidal material. The first method is the easier to understand and is carried out as follows.

A Burton U-tube is employed, consisting of a glass U-tube provided with

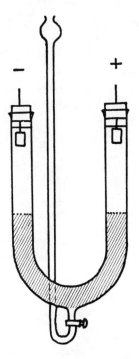

Fig. 39. The Burton tube.

an inlet tube sealed into the middle, as shown in Fig. 39. Water is first poured into the tube, to a height about halfway up either limb. The colloidal suspension is then poured into the thistle funnel and the tap opened slowly, so as to displace the water upwards very gradually, forming a sharp boundary between the clear water and the opalescent suspension below it. Two platinum electrodes, carried in rubber bungs, are inserted into each limb, as shown, and the electrodes are connected to a source of high potential ($c.$ 200 V). Owing to the migration of all the particles in the electrostatic field, the boundary moves accordingly on either side and the rate of movement is measured with a millimetre scale and stop-clock or with a travelling microscope. Knowing the electrical length of the tube and the viscosity of the liquid, the zeta-potential is calculated as follows. Let AB (Fig. 38) be a charged surface in suspension, CD the associated counter-ions, separated from it by a distance d. When a potential is applied, the layers AB and CD separate, one moving to the cathode, the other to the anode. Let the intensity of the electrostatic field be E, the charge density σ, the viscosity of the liquid η and the rate of migration v. Then from Newton's law of viscosity (see Chapter 6) the velocity gradient produced by the motion is v/d, and the frictional force per unit area is therefore $\eta v/d$. But when moving with constant velocity, the frictional and electrical forces must be equal. Hence, from electrostatics, the electrical force per unit area is σE, and so

$$\sigma E = \eta v/d \qquad (2)$$

Substituting for d from eqn (1), it follows that:

$$\zeta = \frac{\eta v}{\epsilon \epsilon_0 E} \qquad (3)$$

Equation (3) is known as the Helmholtz–Smoluchowski equation. Since E, the electrostatic field, is equal to the potential difference divided by the

Table 18

Nature of clay	Zeta-potential (mV)
Ca-clay	− 10
H-clay	− 20
Mg-clay	− 40
Na-clay	− 80
Natural clay	− 30
Treated with 'Calgon'	−135

electrical length of the tube, and η, v and $\epsilon \epsilon_0$ are all known, this equation enables ζ to be calculated.

Values of zeta-potential for a kaolinite with various counter-ions, determined by the author and his associates, are shown in Table 18. These values are in broad agreement with those obtained by other workers. It is generally accepted that for a suspension to be stable, the zeta-potential must be above about 50 mV; hence, the natural clay, which contained a mixture of H, Ca, Mg, Na and K as counter-ions, was in an unstable or flocculated state, as is the case with the majority of natural kaolinites. The very high value obtained with 'Calgon' (sodium hexametaphosphate) is referred to later.

CATION EXCHANGE REACTIONS

If a clay is placed in a solution of a given electrolyte, an exchange occurs between the ions of the clay and those of the electrolyte:

$$X\text{-clay} + Y^+ \rightleftharpoons Y\text{-clay} + X^+$$

As indicated, the reaction is a balanced one and the extent to which the reaction proceeds from left to right depends on the nature of the ions X and Y, their relative concentrations, the nature of the clay, and on any secondary reactions.

Even for equivalent concentrations, some cations are adsorbed more strongly than others; if the cations are written down in order of the readiness with which they are adsorbed, one obtains the *lyotropic* or *Hofmeister series*:

$$H > Al > Ba > Sr > Ca > Mg > NH_4 > K > Na > Li$$

An analogous series can be written for anions, but the latter are more applicable to positive colloids. Comparison with Table 17 shows that, for the alkalis, this is the order of decreasing hydration.

Various theories of the mechanism by which cation exchange takes place have been proposed from time to time. Earlier work (e.g. Wiegner and Jenny, 1927) was based on the Freundlich adsorption isotherm and was of the type:

$$\frac{x}{m} = k\left[\frac{c}{a-c}\right]^{1/p}$$

where x/m = weight of ion adsorbed per gram of adsorbent, c = concentration of ion in solution at equilibrium, and a = amount of electrolyte added. Boyd et al. (1947) adapted the Langmuir isotherm equation, deriving the expression:

$$\frac{x}{m} = \frac{kb_1 c_A}{1 + b_1 c_A + b_2 c_B}$$

for the replacement of a cation B by another cation A, where c_A and c_B are the equilibrium concentrations of A and B respectively, k, b_1 and b_2 are constants and x and m are as above. A kinetic approach was used by Jenny (1936) in deriving his equation:

$$w^2[1 - (V_y/V_x)] - w[S + N] + SN = 0$$

where X and Y are the original and replacing cations respectively, N = amount of Y added, w = amount of exchange that occurs, S = cation exchange capacity, and V_x and V_y are constants for the respective cations. If w is plotted against N for this equation, a hyperbola is obtained which approaches asymptotically the value S as N is increased indefinitely, i.e. complete saturation with Y up to the cation exchange capacity. Gapon (1933) also used a kinetic approach in his equation and this has been investigated by Davidov and Levitskii (1950). Some of the foregoing models show agreement with experimental values over a limited range of concentrations, but no universally applicable equation has been formulated. Many equations that are satisfactory for exchange between cations of like valency break down when applied to unlike cations. Aluminium, which can fit into the kaolinite or montmorillonite structures, is a particularly difficult ion to replace.

More recent work has been concerned with the thermodynamics of cation exchange. Laudelout et al. (1968) investigated the free energy and enthalpy changes for the mutual replacement of Mg, Ca, Sr, Ba, Na and NH_4 in a montmorillonite. Their results showed that the replacement of a divalent cation by one of higher polarisability resulted in a loss of entropy. The enthalpy changes could also be related to polarisability.

Secondary reactions can of course dominate the equilibrium of any exchange reaction. If the ion being displaced, for example, is removed continuously from solution by chemical precipitation or sequestration, the reaction may be virtually complete. If sodium carbonate is reacted with a calcium clay, the Ca ions are removed from solution as insoluble calcium carbonate and the clay is almost completely converted to a sodium clay:

$$\text{Ca-clay} + Na_2CO_3 \rightarrow \text{Na-clay} + CaCO_3$$

This is frequently used as a means of deflocculating a calcium clay, but there are in fact more effective deflocculants, as will be seen later. In the preparation of mono-ionic clays (i.e. clays containing one type of exchangeable cation only) for experimental purposes, the presence of a precipitate may be undesirable; consequently, the method of preparation may consist either of leaching with an excess of electrolyte, subsequently filtering and washing out the excess, or by reaction with an ion-exchange resin:

$$X[\text{resin}] + Y[\text{clay}] \rightleftharpoons Y[\text{resin}] + X[\text{clay}]$$

Since the exchange capacity of such resins is high (c. 250–500 meq/100 g), a four- or fivefold excess only is required for the reaction to be almost complete. The resin is afterwards separated from the clay by sieving. Alternatively, where practicable, a dilute suspension of the clay in water may be passed through a column of the resin; this method is more efficient but cannot be used for clays containing particles much larger than 1 μm.

Cause of cation exchange

In the kaolin minerals, the oxygen and hydroxyl valencies at the planar surfaces of the structure are completely satisfied. At the edges, however, there are aluminium, silicon, oxygen and hydroxyl ions that are not so satisfied, because the lattice is capable of extension indefinitely in the ab plane. These unsatisfied valencies, or 'broken bonds' as they are often called, are satisfied in practice by external ions that do not form part of the structure, but merely act as counter-ions, preserving electrical neutrality. These counter-ions, particularly the cations, are capable of being exchanged for other ions and are one possible cause of cation exchange in clay minerals. 'Broken bonds', however, are not the only cause of cation exchange. In the disordered kaolinites, as mentioned previously, additional balancing cations are present because of the lattice substitutions. These additional cations probably account for the greater part of the cation exchange that occurs with the disordered kaolinites.

Another possible cause of cation exchange in clays, often quoted in the past, is ionisation of basal hydroxyl groups, to produce a negative charge on the oxygen and a hydrogen ion that is exchangeable for other cations. If this were so, one would expect the cation exchange capacity to be strongly dependent on pH, which is certainly not the case. Ionisation of hydroxyl groups is of course an important factor in oxides.

The cation exchange capacity (c.e.c.)

For a given clay, the maximum amount of any one cation that can be taken up is constant and is known as the *cation exchange capacity*, often abbreviated to c.e.c. In principle, the c.e.c. is determined by leaching the clay with a chosen electrolyte, so as to replace all existing cations by one particular cation; the clay is then filtered, washed free of excess electrolyte (often with alcohol rather than water, to avoid hydrolysis) and the amount of chosen cation determined. Ammonium acetate is the electrolyte frequently chosen for this purpose, since ammonium can readily be determined by distillation. Instead of an electrolyte, an exchange resin in the ammonium form may be employed. Other electrolytes, in which the relevant cation can be readily determined chemically, have also been used, such as acetates, sulphates or chlorides of manganese, lithium and sodium. On the whole, values of c.e.c. obtained with different monovalent cations agree reasonably well, but discrepancies are found when comparing monovalent with polyvalent cations. This discrepancy has been ascribed to the formation of complex ions of type —M—OH by polyvalent ions; with calcium, for example, it is conceivable that one monovalent $[Ca-OH]^+$ ion could be attached to every exchange site, resulting in an apparent c.e.c. of twice the normal value if the latter is calculated as Ca^{2+} rather than Ca—OH. Alcoholic solutions of polyvalent cations are said to give normal values of c.e.c., presumably because the complex ions cannot be formed in alcohol.

Certain ions, as we have seen, are prone to be adsorbed very strongly by clay surfaces; if Al or Fe is present as an exchangeable ion, some difficulty may be found in replacing it, consequently giving low c.e.c. values. Again, it is said that calcium is difficult to remove from fuller's earth. Fixation of cations in general may occur when a clay is dried at quite a modest temperature (as low as 100°C); this is particularly the case with potassium montmorillonites, where the potassium ion is situated between the silica sheets and 'fits' tightly in the structure when the silica layers approach closely on drying. Thus, fixation is closely related to ionic size.

Values of cation exchange capacity

For well-crystallised kaolinites (e.g. china clay) the c.e.c. is small, being approximately 2–5 meq/100 g. The amount of c.e.c. contributed by broken bonds is probably small, since the crystals are relatively large; moreover, the degree of substitution is small. For the disordered kaolinites, however, the c.e.c. is high, and of the order of 30–40 meq/100 g, due to lattice substitution. Some authorities believe that the c.e.c. of well-crystallised kaolinites is due to substitution rather than broken bonds, but the degree of substitution required is so minute that it would be difficult to detect by chemical analysis. Many workers have reported that c.e.c. increases with specific surface area (or as particle size decreases). This would indeed be true if c.e.c. were associated entirely with broken bonds at the edges, and many authorities have found a linear relationship between c.e.c. and surface area for well-crystallised kaolinites. However, rather different results were obtained by the author for disordered kaolinites (fireclay). In this investigation each clay was divided by centrifugal sedimentation into a number of fractions, each containing particles of a specified size. Above about 0·3 μm, the fractions are contaminated with coarse-grained quartz, the c.e.c. of which is negligible. As the particle size decreases, the amount of quartz decreases, so that the c.e.c. at first rises as shown in Fig. 40. Below about 0·3 μm, however, quartz is finally eliminated, so that the fractions below this size are almost pure clay mineral. As will be seen from Fig. 40, the c.e.c. of the pure clay mineral is very nearly independent of particle size. Very recent work has shown that whilst in some instances a slight variation with particle size can still be detected below 0·3 μm, this is

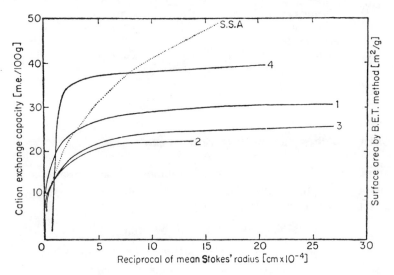

Fig. 40. Particle size and cation exchange.

due to variations in crystallinity (and probably substitution also) rather than particle size as such.

The suggestion has sometimes been made that the high c.e.c. of fireclays and ball clays is due, not to disordered kaolinite, but to illite. That this is not so is shown clearly by Fig. 40, where clay 1, a bond clay, was completely devoid of alkali-bearing minerals. Similar results have since been obtained with a white-burning, disordered kaolinite found in Jamaica. Where reasonably pure 'illite' can be isolated, it appears to have a c.e.c. similar to that of disordered kaolinite. Typical c.e.c.s for different naturally occurring clays are shown in Table 19.

Table 19
Typical Values of Cation Exchange Capacity for Various Clays

Clay	Cation exchange capacity (meq/100 g)
Coal Measure fireclay	7–12
Ball clay	10–20
China clay	2–5
Brick clay	7
Fuller's earth	60
Bentonite	80–120
Halloysite	13
Vermiculite	100–150

Cation exchange reactions with organic ions

As might be expected, basic organic compounds that ionise in aqueous solution may also replace other cations on clay surfaces. Amines, for example, may react with calcium clays:

$$R.NH_3^+ + \text{Ca-clay} \rightarrow R.NH_3\text{clay} + Ca^{2+}$$

In the above equation, R stands for an alkyl or aryl group. Whatever the nature of the original counter-ions, replacement of the latter by an amine invariably results, in aqueous suspensions, in the clay being strongly flocculated, possibly because the amine is strongly attracted to the Stern layer. Another explanation may be that since the amine is adsorbed with the NH_3^+ group close to the surface, the alkyl or aryl group R projecting outwards, the surface presented to the suspending medium is effectively an 'organic' surface and is therefore hydrophobic; thus, the solvation energy of the composite clay particle is drastically reduced. Conversely, if the amine-clad clay is dispersed in an organic medium, the solvation energy is very high and this alone may account for the deflocculation that occurs.

Because of their *organophilic* properties, amine-clad clays have many industrial applications. Their compatibility with organic media enables them to be used as paint-thickeners, in polishes, and in treatment of effluents.

ANION EXCHANGE

Earlier workers on ion exchange in clay soils found that the alkalinity of NaOH and other bases was reduced when they were placed in contact with such soils; this was then attributed to the adsorption of hydroxyl ions by the clay and it was therefore supposed that other anions could be adsorbed and so exchanged. It has since been shown that this view is incorrect; all natural clays contain a proportion of exchangeable hydrogen, and it is the latter which is responsible for the reduction of alkalinity, by direct cation exchange followed by formation of water:

$$\text{H-clay} + \text{NaOH} \rightleftharpoons \text{Na-clay} + \text{H}_2\text{O}$$

This reaction, as shown, is reversible and thus a sodium clay is slightly hydrolysed in water.

Again, claims that kaolinitic soils adsorbed such anions as phosphate and fluoride were shown to be erroneous, subsequent work having shown that the alleged adsorption was caused by direct chemical reaction of phosphate with iron compounds in the soil, rather than by reaction with the clay.

However, a limited amount of anion exchange might be expected to occur at the edges of clay particles. The 'broken bonds' at the edges, mentioned in connection with cation exchange, may give rise to either positive or negative sites, depending on whether the site is opposite O^{2-} ions or opposite Al^{3+} or Si^{4+}. These sites would normally be occupied by OH_3^+ or OH^- respectively, acting as counter-ions. The extent to which these ions are replaceable obviously depends on the pH, as with all amphoteric surfaces; according to some authorities, the hydroxyl ions ionise at a pH below 6·5, but others assert that the edges can remain positive, due to hydroxyl ionisation, at pH values well above 7. Whichever view is correct, there seems no doubt that the edges of clay particles are frequently positive when in their natural state, i.e. no chemical dispersants having been added.

A limited amount of anion exchange does appear to take place with kaolinites, but the anion exchange capacity is less than one-tenth the cation exchange capacity. Moreover, it is likely that with large polyanions, physical adsorption may occur in addition to anion exchange.

Despite the amphoteric character of the crystal edges, it should be realised that the clay as a whole remains predominantly negative, owing to the inherent lattice charge, which outweighs any positive charge that may develop locally.

DEFLOCCULATION AND FLOCCULATION OF CLAYS

The kaolin group

As explained previously, the stability of a lyophobic suspension depends primarily on the nature and concentration of the counter-ions; for kaolin-group minerals, sodium ions favour stability and complete deflocculation, whilst hydrogen and calcium ions cause flocculation to occur. Deflocculated kaolinites are required for casting slips and in the determination of particle size; flocculated kaolinites, on the other hand, may be required in various dewatering processes such as filter pressing. In general, deflocculated clays have a characteristically lower viscosity than flocculated clays; this subject is discussed more fully in Chapter 6.

One method of achieving deflocculation has been mentioned, namely, the precipitation and replacement mechanism, in which the displaced ion is removed from solution by precipitation or sequestration. Thus, calcium clays may be deflocculated with sodium carbonate, sodium oxalate or sodium phosphate; hydrogen clays may be deflocculated with sodium hydroxide. There is a class of deflocculants, however, that appears to function differently. Sodium silicate, for instance, is a very powerful deflocculant for the majority of natural clays. Contrary to what one might expect, the mechanism in this instance is not simply the replacement of other ions by sodium, followed by precipitation of the displaced cations as silicates. Simple measurements have shown that a high zeta-potential, accompanied by deflocculation, can be produced when only a fraction of the total cation sites have been replaced by sodium. Similar results are observed with other polyelectrolytes such as 'Calgon' (sodium hexametaphosphate) and 'Dispex' (sodium polyacrylate). These observations suggest that the anion plays an important part in deflocculation. Subsequent measurements have confirmed that this is so; large polyanions such as silicate, polyphosphate and polyacrylate are adsorbed by clay surfaces, in addition to sodium ions. It is not quite clear how this adsorption produces a high zeta-potential, but it would seem likely that the adsorbed polyanions provide extra negative sites that are strongly ionised, so that the associated cations are at a relatively great distance from the surface. The polyanions are likely to be repelled by the negative planar faces of clay crystals, and it is therefore probable that they are adsorbed on to the edges, either by anion exchange or by physical adsorption. Another characteristic of polyelectrolyte deflocculants is that a considerable excess can be tolerated without offsetting the deflocculating effect.

The effect of edge charges

As was mentioned earlier, the edges of clay crystals may behave differently from the planar surfaces in having an amphoteric character. It is clearly possible for the edges to carry a net positive charge over a wide range of pH, whilst the faces are permanently negative. In these circumstances, the edges and faces will mutually attract, giving rise to a so-called 'edge-to-edge' flocculation, even though the zeta-potential may be moderately high.

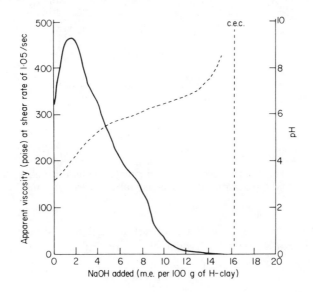

Fig. 41. pH and viscosity curves for hydrogen ball clay. Solid line = viscosity. Broken line = pH.

Because of the comparatively small magnitude of the edge charges, they can be 'neutralised' or reversed by small additions of a suitable electrolyte.

It has been observed that when a hydrogen clay is treated with successive small additions of NaOH, the degree of flocculation, and therefore the viscosity, at first begins to increase markedly until just before the equivalence point, after which deflocculation suddenly occurs and the viscosity drops sharply (Fig. 41). This is because the sodium ions first replace H^+ on the faces only, increasing their negative potential without influencing the edges; this results in an increased attraction (initially) between edges and faces. On further addition of NaOH, the pH rises sufficiently to reverse the edge charges, so that the clay particles become negatively charged overall and the system is therefore deflocculated (Fig. 42).

It has been suggested that the remarkably small additions of polyelectrolyte required to deflocculate clays can be explained by supposing that the polyanions are adsorbed on to the crystal edges and thus 'mask' any positive charge on these sites. This explanation cannot be regarded as

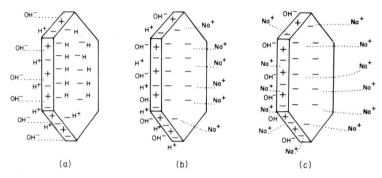

Fig. 42. Edge and face charges on (a) H-clay (acid environment); (b) Na-clay (acid environment); (c) Na-clay (alkaline environment).

satisfactory, however, since it does not account for the very high zeta-potential that is observed when clays are deflocculated in this way (Table 18).

Other causes of flocculation

Natural clays, where the predominant exchangeable cation is calcium, are normally flocculated. The degree of flocculation is in all cases increased by the addition of any electrolyte in sufficient quantity, due to 'crowding' of the Stern layer. Salts of polyvalent cations are more effective flocculants than those of monovalent cations. Thus, the deflocculation of a natural clay may be inhibited by the presence of soluble salts, notably sulphates of calcium, magnesium or iron. If the processing of the clay involves filter-pressing, much of the soluble salt will thereby be removed; otherwise, careful addition of the stoichiometric amount of barium carbonate may effectively remove the sulphates:

$$CaSO_4 + BaCO_3 \rightarrow CaCO_3 + BaSO_4$$

The net effect of this treatment, as indicated by the equation, is to produce two sparingly soluble substances, the effect of which is thus minimised. Deflocculation can often subsequently be achieved by the addition of sodium carbonate, sodium silicate or both.

Sometimes it is required to flocculate a clay to enable it to be dewatered: for example, in the treatment of coal 'fines' contaminated with clay. For this purpose, some organic flocculants such as polyacrylamides may be more effective than electrolytes. These organic flocculants may act in several ways: by entering the Stern layer, by rendering the surface of the clay hydrophobic, or by linking clay particles together via long-chain molecules.

Sedimentation volume

If a suspension of clay is allowed to stand for a sufficiently long time, the particles gradually fall out of suspension, forming a *sediment* at the bottom. A deflocculated suspension requires a considerable time for sedimentation (the supernatant liquid may never completely clear) whilst a flocculated suspension may sediment completely in half an hour or less, depending on the depth of liquid and the degree of flocculation.

The nature of the sediment gives a useful insight into the forces between the particles and the type of aggregation. A strong flocculated suspension contains large, fluffy aggregates of particles, which contain large numbers of voids and so occupy a relatively large volume. A deflocculated suspension, by contrast, consists of largely discrete particles which are able to pack efficiently and occupy a relatively small volume. The volume of sediment produced by 1 g of clay in suspension is therefore a useful index of the degree of dispersion or flocculation.

The smectite group (montmorillonites)

The smectites have a high cation exchange capacity (some 100 m.e. 100 g^{-1}) and would therefore be expected to respond markedly to ion exchange reactions and in particular to deflocculating agents. Although the monovalent ions Li^+, Na^+ and K^+ do produce much higher zeta-potentials than the polyvalent ions Mg^{2+}, Ca^{2+}, Ba^{2+}, etc., there is no clear correspondence with rheological properties. Baver (1929) and Baver and Winterkorn (1935) found that the addition of NaOH to a hydrogen beidellite caused at first an increase in viscosity, due possibly to the edge effects reported by the present author for kaolinite (Fig. 42); on further addition of NaOH the viscosity subsequently fell to just short of its original value, then began to rise again on addition of excess. Divalent bases such as $Ca(OH)_2$ caused an increase in viscosity with no minimum value. Unlike kaolinite, this montmorillonite-type clay was not deflocculated to any extent by the addition of NaOH or any other alkali.

Somewhat similar results were obtained by Zamudio-Castillo (1983), who prepared a series of mono-ionic montmorillonites by resin treatment. For a given consistency, much higher concentrations of Ca, Mg and Al clays were required than of the Li, Na and K clays. These results are summarised in the table below, which shows the concentrations (w/v) of various clays required to produce an equilibrium stress of 80 N m^{-2} at a shear rate of 1914 s^{-1}.

Cation	Na	K	Li	NH$_4$	H	Ca	Mg	Fe	Al
Concentration (w/v)	4·63	4·89	4·26	6·46	4·76	9·06	9·06	7·67	14·26

The situation is thus almost the reverse of that reported for kaolinites in the preceding section; the Li, Na and K smectites tend to be more viscous than the Ca, Mg, Fe and Al clays. In order to understand this very different behaviour, it is necessary to recall that in the smectites, the lattice is readily penetrated by water; in fact, the exchangeable cations, together with their water of hydration, are situated in interlayer positions. Considering the sodium montmorillonite, it is clear that on suspending in water swelling will occur as water enters the structure, thus increasing the effective volume occupied by the solid, which in turn causes an increase in viscosity. Since the zeta-potential on each internal surface is high, due to the sodium ions, the electrostatic repulsion forces are effective at long range, thus increasing further this swelling effect. Eventually a balance is established between attractive and repulsive forces, resulting in a highly viscous structure that may, if undisturbed, produce a gel in which a relatively large proportion of water is entrapped. It is not certain whether this swelling alone is responsible for gelling or whether edge-to-face bonds between particles may also contribute to gel strength. Such bonds should, of course, be eliminated by raising the pH, as described for kaolinite; the findings of Baver (1929), Baver and Winterkorn (1935) and of Zamudio-Castillo (1983), however, show clearly that no abrupt fall in viscosity occurs on the addition of excess NaOH, despite the increase in pH; thus, if edge-to-face bonds do exist in this system, their contribution to gel strength must be minimal. The only exception seems to be the H-clay, the viscosity of which is somewhat reduced by the addition of NaOH; in none of these instances however can the addition of NaOH be said to deflocculate the system.

By contrast, if a Ca, Mg, Fe or Al clay is considered, the zeta potential is so low that forces of attraction between the unit layers predominate, thus keeping swelling to a minimum when the clay is suspended in water. Consequently, the volume effectively occupied by the solid is relatively small, resulting in turn in a much lower viscosity.

It should not be concluded that the smectites cannot be deflocculated at all, though it will be clear that the simple process of cation exchange is not, in general, effective. A reasonable degree of deflocculation can be achieved by the addition of polyacrylates or polyphosphates to which the Li, Na, and K clays respond more readily than do the polyvalent species. The precise mechanism by which smectites are deflocculated by polyelectrolytes is not fully understood; it will be clear that adsorption of the polyanion on crystal edges cannot be the sole cause but it may be that the polyelectrolytes in some way inhibit swelling when adsorbed on interlayer sites.

The above discussion applies, strictly speaking, only to pure smectites; naturally occurring clays, e.g. the bentonites and fuller's earth, contain non-clay impurities which can profoundly modify the flow properties. The author has found that a Wyoming bentonite, if not pre-treated to remove quartz and other impurities, could form a fluid suspension at a concentration as high as 40% (w/v). The effect of *ageing* should also be taken into account; this topic is discussed in Chapter 6.

PHYSICAL ADSORPTION BY CLAYS

The atoms or ions at the surface of a solid are subject to a different net force from those in the interior, because at the surface there are atoms or ions on one side only. Accordingly, surfaces in general have a greater activity than the interior and are prone to attract other molecules or ions; this phenomenon is known as *adsorption*. The effect of such adsorption is to reduce the free energy of the surface.

Cation exchange is a special case of adsorption, which is specific for cations only. Chemisorption is another special case, where atoms or ions of a particular type only are adsorbed by a surface, e.g. the adsorption of oxygen by metal surfaces. *Physical adsorption*, however, is not specific. A wide variety of different non-ionogenic substances can be adsorbed on to surfaces, the forces by which they are held being of van der Waals type.

Whilst for ion exchange it is possible for a reaction to be virtually complete and irreversible, with physical adsorption there is always an equilibrium between adsorbed and unadsorbed substance. For adsorption of solute from a solution, the Freundlich isotherm equation is frequently obeyed:

$$x/m = k \cdot c^{1/n}$$

where x = weight of solute adsorbed, m = weight of absorbent, c = concentration of solute at equilibrium, k = a constant related to the surface area of the absorbent and n = a constant, usually greater than unity. Unlike ion exchange, physical adsorption is markedly dependent on temperature. Another equation that has been applied to physical adsorption from solution is a form of the Langmuir equation, originally derived kinetically for adsorption of gases:

$$x/m = \frac{(x/m)_{max} ac}{1 + ac}$$

where $(x/m)_{max}$ = the monolayer capacity, i.e. the amount of solute required to form a monomolecular layer on the surface of 1 g of solid. It will be clear from this equation that as c increases indefinitely, x/m approaches the limiting value $(x/m)_{max}$. Unlike the Freundlich isotherm therefore, the Langmuir equation is only valid where adsorption is limited to a monolayer. Many instances of this are known, however; e.g. the adsorption of amines by clays from xylene solution.

A third equation of very general applicability is the Gibb's isotherm, which unlike the two previous expressions is based on thermodynamic reasoning. Essentially, it is based on the fact that adsorption from a solution on to a solid surface is favoured when the surface tension between the solid and liquid (i.e. the surface free energy) is reduced by this process. The equation may be written:

$$w = -kAc \cdot \frac{d\gamma}{dc}$$

where w = mass of solute adsorbed, A = surface area of solid adsorbent, γ = surface tension and k = a constant.

The negative sign appears because *positive* adsorption occurs when there is a *reduction* of surface tension, i.e. when $d\gamma/dc$ is negative. One difficulty of applying this equation to experimental data is that $d\gamma/dc$ relates to the surface tension between the solid and the liquid (which is inconvenient to measure), whilst most published data refer to the surface tension between the liquid and air; moreover, the quantity $d\gamma/dc$ may vary with concentration. The equation is no longer valid where other factors such as Coulombic forces are involved in adsorption, as in ion exchange. In the absence of such factors, the advantage of the equation is that it may be applied to a wide range of solvents.

Adsorption of basic dyes by clays

The uptake of certain dyestuffs by clays from aqueous solution was formerly considered to be physical adsorption only. It is now known that with basic dyes at least, cation exchange first occurs, giving way to physical adsorption only when complete replacement of the original ions by basic dye ions has occurred. The adsorption of the basic dye methylene blue has often in the past been used as a measure of cation exchange capacity; at other times it has been used as an index of surface area or associated properties such as plasticity. These conflicting claims are probably a reflection of the fact that both cation exchange and physical adsorption occur, it being difficult to distinguish between the two. Work by the author has shown that the adsorption of methylene blue by clay occurs in three distinct stages. If successive small increments of a 0·01 M solution of the dye is added to a weighed amount of clay (1 g is suitable for kaolinites), methylene blue ions are adsorbed strongly by ion exchange, until the surface is completely covered by a monomolecular layer of dye. This initial layer is adsorbed irreversibly and if the clay–dye solution is allowed to sediment, the supernatant liquid is colourless, showing that all the added dye has been adsorbed. Further additions of dye beyond the monolayer stage are then adsorbed by cation exchange if unexchanged sites are still available, or by physical adsorption; these further adsorption stages are, however, reversible and on sedimentation of the clay the supernatant liquid is distinctly coloured. Finally, when all the cation sites are occupied by dye cations, the process proceeds entirely by physical adsorption. The reversibility of the adsorption beyond the monolayer stage occurs because second and subsequent layers of dye cannot be subject to as strong a coulombic force as the initial layer.

Although, as the author has shown, it is quite possible to measure c.e.c. if a large excess of dye is used, the method is less convenient than that using ammonium acetate. The limit of irreversible adsorption is readily determined, however, and is clearly the most useful measurement, since from the amount of dye required to form a monolayer, the specific surface area of the clay can be calculated, provided the effective area covered per molecule is known or measured. In practice, a number of 1 g samples of clay are shaken with some 10 ml of water in separate test-tubes. To the first tube is added 1 ml of 0·01 M methylene blue solution, to the second 2 ml and so on. After thorough shaking, the tubes are allowed to stand until the clay is sedimented to form a more or less clear supernatant liquid, free from clay particles. Generally, the supernatant liquid in the first one or two tubes is colourless, but with increasing dye additions a stage can be recognised where the supernatant first begins to show a colour; the amount of dye added at this stage is the limit of irreversible adsorption (L.I.A.), corresponding to the formation of a monomolecular layer of dye. Beyond this, the solutions are seen to be more and more deeply coloured due to reversible adsorption.

By plotting the L.I.A. against surface area determined by gas adsorption, the author found that there was a linear relationship for all clays tested; from the slope of the graphs, the average area occupied per dye molecule was found to be 26 $Å^2$. Using this value, the specific surface areas of other clays can be determined as described. The value of 26 $Å^2$ for the area of the dye molecule indicates that the latter is adsorbed 'end-on', so that this value is the minimum possible for methylene blue. With different types of clay and under different conditions, it is of course possible that the dye molecules adsorb with a different orientation. There is some evidence that when methylene blue is adsorbed between the silica sheets of a montmorillonite, the effective area covered per molecule is much greater than 26 $Å^2$ because the molecule lies flat on the surface. This variation of area covered is clearly a possible source of error in surface area measurements, but provided the method is calibrated for a particular clay under standard conditions, the error should be minimised. A further source of error may occur with clays having a low c.e.c., in which case there may be insufficient exchange sites to form a monolayer. This could well occur with some well-crystallised kaolinites and in such instances a different dye of larger molecular area might give better results. Other basic dyes that behave in a similar manner to methylene blue are Bismarck brown, safranine, methyl violet and crystal violet. In selecting a dye for routine work, it is advisable to check first upon its toxicity and possible carcinogenic properties; a number of dyes containing condensed aromatic rings are known to be carcinogenic.

Adsorption of acid dyes

As the name suggests, acid dyes ionise in solution to form an inorganic cation, usually Na, and a complex organic anion, the latter being coloured. In the presence of a clay mineral, Na ions will clearly displace other cations, but the coloured component, the anion, can only be adsorbed physically or by anion exchange. Physical adsorption appears to take place most readily if the ionisation of the dye is suppressed, for example by the addition of HCl to lower the pH. However, the colour of many acid dyes changes with pH and consequently measurements of physical adsorption are not a promising means of determining surface area. Moreover, with physical adsorption, calibration is difficult because of the two 'unknowns', n and k, in the Freundlich equation. It has been observed that a minute proportion of acid dye is irreversibly adsorbed, but since this does not appear to be related to surface area, the process is probably anion exchange at crystal edges.

Adsorption from non-aqueous solvents

For the measurement of surface area by dye adsorption, the most suitable system would appear to be a non-ionogenic dye dissolved in a non-ionising solvent, since this would avoid complications due to ion exchange and possibly the adsorption of solvent by the solid surface. It has been found that clays and other ceramic materials do indeed adsorb non-ionogenic substances from non-aqueous solvents, so that in such instances the adsorption can only be non-specific, i.e. physical adsorption. One such system that has been investigated is the adsorption of the dye dimethyl yellow from solution in xylene, for which the Gibb's equation was obeyed for adsorption not exceeding a monolayer. For the measurement of surface area, provided $d\gamma/dc$ is constant over the experimental range of concentration c, the equation simplifies to:

$$w = k' A c$$

where k' replaces $k \cdot d\gamma/dc$. Then, if the adsorption w is measured for several different values of concentration c, a graph of w against c produces a straight line of gradient $k'A$. Adsorption measurements of dimethyl yellow in xylene can be used for measuring the surface areas of various substances, since the adsorption does not depend on the presence of ion-exchange sites. Prior calibration against a standard (gas adsorption) method is still necessary, but if w is plotted against c for various values of the latter, the slope of the linear plot is proportional to surface area. In this way, values of surface area for calcined alumina, flint, bone and ceramic colouring agents can be obtained. However, separate calibration is necessary for each type of substance being studied, implying that the value of k' in the equation is not constant or that $d\gamma/dc$ is not constant.

Orientation of adsorbed molecules

There is another aspect of orientation of adsorbed substances which is important in determining compatibility with the liquid medium. Where no strong preference occurs for a particular mode of adsorption, a *normal* orientation is taken up, i.e. lyophilic groups of the molecule project outwards into the liquid, and lyophobic parts of the molecule are attached to the solid surface. Thus, stearic acid in water takes up a normal orientation with respect to a clay surface, with the hydrophilic —COOH groups projecting outwards so as to be in contact with the water, and the lyophobic group attached to the clay. In a non-aqueous medium, the opposite orientation would be *normal*, because the carbon chain is then attracted by the solvent and the —COOH group is nearest to the clay surface. The orientation can be *inverse* when a group is present that has a strong attraction for the clay surface. An example is an amine, which ionises to give a positive ion, so that the charged part of the molecule, the —NH_3^+ group, is attached to the clay surface, and the carbon chain projects outwards. This constitutes *inverse orientation* in water, since the carbon chain is hydrophobic. The adsorbed amine would, of course, have a *normal* orientation in a non-aqueous medium, since the projecting carbon chain is solvent-attracting.

Adsorption of gases and liquids by clays

In dealing with colloidal systems it was pointed out that because of its polar character, water could be adsorbed by clay surfaces. Kaolin-type clays adsorb 1 or 2% by weight of water from the atmosphere at normal temperatures, but this adsorbed water is readily desorbed by drying the clay at or above 105°C. Measurement of water adsorption at a controlled humidity is an approximate measure of the specific surface area of a clay and has been used for its characterisation (see Keeling, 1961; Chapter 9). Direct calculation of surface areas from water adsorption is not possible because of the uncertainty in the number of adsorbed layers formed.

Montmorillonites adsorb water to a much greater extent than kaolinites because water is adsorbed not only on the external surfaces but is able to penetrate between the silica sheets, causing the basal spacing to increase and resulting in a visible swelling of the clay. Polar molecules in general are able to force the silica layers apart because of the relatively weak bonding between them; glycerol, for example, expands the basal spacing to a constant value of 17 Å, a fact made use of in the identification of montmorillonites. The amount of water adsorbed by montmorillonites depends on the nature of the clay, the humidity, the temperature and the nature of the exchangeable cations which are situated between the silica layers. Because the exchangeable cations themselves are considered to be hydrated, it is difficult to distinguish between water held by the surfaces and that bound to exchangeable cations. When montmorillonites containing adsorbed water are heated, the temperature being raised gradually, desorption of water occurs but some is not desorbed until the temperature is raised to about 350°C or above, and accordingly naturally occurring montmorillonites invariably contain a proportion of adsorbed water.

Dehydration studies, for example thermal analysis, show that the adsorbed water may be evolved in two or more definite stages, depending on the exchangeable cations. One stage in the dehydration is considered to be the removal of water from the clay surface, a further stage the dehydration of the cation. A third stage, if observed, is attributed to the presence of more than one layer of water molecules attached to the cation. Some of the conclusions drawn about cation hydration may be open to question but it is nevertheless interesting that estimates of the degree of hydration of various cations from thermal measurements frequently disagree with those quoted from other sources (Table 17).

It has been supposed that, for all clay surfaces, the degree of orientation of the adsorbed water molecules diminishes progressively with distance from the surface, until a point is reached where the thermal motion of the molecules overcomes the electrostatic force and the molecules cease to be attached to the surface. Both the direction and the extent of orientation are of course influenced strongly by the exchangeable cations.

There is considerable disagreement about the maximum probable thickness of the adsorbed water on clay surfaces. It has been estimated that the coulombic forces at the clay surface are not likely to extend the adsorbed water film beyond the thickness of about 10 Å, whilst experiments on water adsorption have yielded values around 15 Å for vapour phase adsorption. Few measurements seem to have been made of adsorption from the liquid phase. Davis and Worrall (1971) found that when clays were equilibrated with a dilute solution of glucose, the concentration of the latter apparently increased, which is consistent with adsorption of the liquid (water) rather than the solute. Calculations based on these results gave values of a few hundred Å for the thickness of the adsorbed layer, depending on the nature of the exchangeable cation. These high values, however, probably indicate the thickness of 'structured' or orientated water, rather than the extent of firmly bound water. This is supported by recent work on the structure of water which suggests that some degree of loose structure can be detected at a few hundred Å from a hydrated cation, there being a likelihood that much of this structure would be broken down by shear when the cation is made to migrate.

Similar high values have been calculated from the Atterberg lower liquid limit (see later), assuming that all the water at this concentration is bound to the surface; but it is probable under these conditions that much of the water is not truly adsorbed but merely entrapped in a 'network' of linked clay particles, similar to the so-called 'scaffold structure' associated with thixotropic systems.

Many other polar liquids, particularly alcohols, are adsorbed by clays. The adsorption by clays of inert gases at low temperatures and pressures forms the basis of the well-known B.E.T. method for the determination of surface area.

INTERCALATION COMPOUNDS

Until recently, it has been considered that the kaolinite structure, in contrast to montmorillonite, was not capable of being expanded. Whilst this is true for the conditions under which clays are normally utilised, it is now clear that under rather more drastic conditions the hydrogen bonding between units can be overcome. For example, concentrated solutions of potassium acetate, urea, formamide and hydrazine, when reacted with kaolinite, produce intercalation compounds having a basal spacing of 10–11 Å as compared with 7·14 Å for unexpanded kaolinite. It is believed that with these compounds hydrogen bonding occurs between the OH surfaces of the kaolinite and the —CO or NH_2 groups of the intercalating substances.

These intercalation compounds do not appear to be particularly stable, since addition of excess water dissolves out the reagent and the kaolinite lattice contracts to its normal basal spacing.

Research by Weiss et al. (1966) established that intercalation complexes with kaolinite could be formed by direct reaction of the mineral with a concentrated aqueous solution of the reagent, with the reagent alone if a liquid, or with the molten solid. Some intercalating reagents do not react directly with kaolinite but are capable of displacing another complexing agent.

Wiewora and Brindley (1969) found that intercalation in kaolinites was seldom complete; some disordered kaolinites proved to be almost completely unreactive in this respect. No satisfactory explanation has yet been put forward for this rather surprising behaviour. Since, however, the disordered kaolinites in general have a high cation exchange capacity, it is possible that the relatively large concentration of exchangeable cations in external sites around the crystallites or as inclusions in lattice defects may prevent ready access of the intercalating agent.

The expansion of the basal spacings of various minerals by intercalating agents has been suggested as a means of identification, though it would be inadvisable to rely on this method alone. It appears that of the other kaolin minerals, halloysite intercalates more readily than does kaolinite, dickite rather less so.

The ready intercalation of the montmorillonites by water has already been referred to but the relatively open structure of these minerals favours intercalation also by glycerol, ethylene glycol and a number of other compounds. The complexes formed between montmorillonite and glycerol or ethylene glycol have been utilised in the identification of this mineral (see Chapter 9).

HYDROGEN CLAYS AND pH TITRATION CURVES

Although the hydroxonium ion (OH_3^+) can readily replace other cations associated with clay, the preparation of hydrogen clays is complicated by secondary reactions. Kaolin-type clays are not appreciably attacked by dilute HCl at ordinary temperatures, but montmorillonites are said to be much more susceptible.

An alternative method that has been used is that of *electrodialysis*. In this procedure, an aqueous suspension of the clay is placed in a porous pot, into which dips a platinum foil electrode. The porous pot is immersed in a larger glass vessel containing distilled water, in which is placed another platinum foil electrode. A potential of some 200 V is applied to the electrodes, so that the one in contact with the clay is positive. The principle is similar to electrophoresis, but since the clay particles are prevented from migrating by the porous barrier, the main reaction is the migration of exchangeable cations to the cathode.

The reaction requires some 24 hr for completion and results in the exchangeable cations appearing at the cathode as hydroxides, with the formation of hydrogen clay, so that the reaction may be written:

$$\text{X-clay} + \text{H}_2\text{O} \rightarrow \text{H-clay} + \text{XOH}$$

where X represents the original exchangeable cations. Electrodialysis is rarely complete in one operation, however; the treated clay frequently has to be removed and electrodialysed once or twice more with fresh distilled water.

Although aqueous acids are avoided in electrodialysis, some decomposition of montmorillonites nevertheless occurs, whilst kaolinites are but little affected. Another and possibly better method is treatment of the clay with a cation exchange resin in the hydrogen form:

$$\text{H-resin} + \text{X-clay} \rightarrow \text{X-resin} + \text{H-clay}$$

The use of acid is avoided and there is evidence that even montmorillonites are not seriously affected by this treatment.

Many authorities are nevertheless of the opinion that hydrogen clays are inherently unstable and have adduced evidence that the low surface pH due to the hydrogen ions causes the clay surface to be attacked, with the result that aluminium ions displace hydrogen from the exchange sites, forming an aluminium clay.

Despite their alleged instability, hydrogen kaolinites certainly exist and have been found to occur naturally. Moreover, it can readily be shown that they behave like weak acids, releasing a small proportion of exchangeable hydrogen ions in ionised form, so that the pH of their aqueous suspensions may be 3 or lower. When titrated against NaOH, an acid–base reaction occurs with the formation of the sodium clay:

$$\text{H-clay} + \text{NaOH} = \text{Na-clay} + \text{H}_2\text{O}$$

If the titration is done potentiometrically, plotting pH against amount of NaOH added, a curve of the type shown in Fig. 43 is obtained. The initial pH is quite low, as will be seen, there being a definite inflection at the equivalence point, corresponding to the cation exchange capacity of the clay. The equivalence point occurs at pH 7·5–8 for kaolinite, indicating that the Na-clay formed hydrolyses slightly.

It has been observed that the inflection is made sharper if a strong

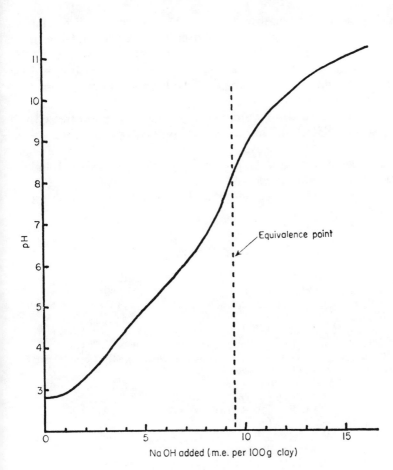

Fig. 43. A pH titration curve.

electrolyte that does not interfere with the reaction, e.g. KCl, is added before titration. The reason for this is obscure, but it may be that it is due to the expulsion of hydrogen ions by KCl by normal ion exchange, enabling equilibrium to be attained more quickly between additions of NaOH. Alternatively, it may be that the added electrolyte assists in the removal of amorphous alumina and silica that may 'clog' the exchange sites.

If the titration is carried out with a natural clay, two inflections are sometimes obtained. This is because the reaction then proceeds in two stages: (a) the ready replacement of OH_3 ions by Na; (b) the more difficult replacement of Ca ions by Na. It has been suggested that a further inflection in the titration curve could be produced by replacement of hydrogen in OH groups at high pH values, but there is no evidence for this with kaolinite titration curves. Similar results have been obtained for montmorillonite, but the instability of the hydrogen form renders them open to doubt.

As an alternative to direct titration, a hydrogen clay may be treated with an excess of NaOH and the excess base titrated against standard acid. Using this procedure for a hydrogen kaolinite, the author obtained results in close agreement with those obtained by direct titration. Thus, the reduction in alkalinity is due solely to the combination of OH_3 and OH to form water and not to direct adsorption of OH ions by the clay surface, as has often been claimed.

REFERENCES

BAVER, L. D., *Univ. Missouri Agr. Exptl. Sta. Res. Bull.*, **129** (1929).
BAVER, L. D., and WINTERKORN, H. F., *Soil Sci.*, **40** (1935) 403.
BOYD, G. E., SCHUBERT, J., and ADAMSON, A. W., *J. Am. Ceram. Soc.*, **69** (1947) 2818.
DAVIDOV, A. T., and LEVITSKII, I. Y., *Zhur. Obschei. Khim.*, **20** (1950) 1776.
DAVIS, G. A., and WORRALL, W. E., *Trans. Brit. Ceram. Soc.*, **70** (1971) 71.
DERYAGIN, B. V., and LANDAU, L., *Acta. Phys. Chim.*, *URSS*, **14** (1941) 633.
GAPON, E. N., *Zhur. Obschei. Khim.*, **3** (1933) 144.
JENNY, H., *Phys. Chem.*, **40** (1936) 501.
KEELING, P. S., *Trans. Brit. Ceram. Soc.*, **60** (1961) 217.
LAUDELOUT, H., VAN BLADEL, R., BOLT, G. H., and PAGE, A. L., *Trans. Farad. Soc.*, **64**(5) (1968) 1477.
VERWEY, E. J. W., and OVERBEEK, J. Th. G., *Theory of the Stability of Lyophobic Colloids*, Elsevier Science Publishers, 1948.
WEISS, A., RANGE, K. J., LECHNER, H., and THIELEPAPE, W., *Proc. Int. Clay Conf.*, Jerusalem, **2** (1966) 8.
WIEGNER, G., and JENNY, H., *Kolloid-Z.*, **43** (1927) 268.
WIEWORA, A., and BRINDLEY, G. W., *Proc. Int. Clay Conf.*, Tokyo, **1** (1969) 723.
ZAMUDIO-CASTILLO, M. R., PhD Thesis, University of Leeds, 1983.

READING LIST

S. GLASSTONE, *Textbook of Physical Chemistry*, Macmillan, 1956.
R. E. GRIM, *Clay Mineralogy*, McGraw-Hill, 1958.
C. E. MARSHALL, *The Colloid Chemistry of the Silicate Minerals*, Academic Press, 1949.
D. J. SHAW, *Introduction to Colloid and Surface Chemistry*, Butterworth, 1970.
H. VAN OLPHEN, *An Introduction to Clay Colloid Chemistry*, John Wiley, 1977.

Chapter 6

The Rheology of Clay—Water Systems

THE IMPORTANCE OF RHEOLOGICAL STUDIES

Rheology is a broad term meaning the study of flow and deformation of matter. The way in which clay–water systems flow under stress is of very great importance in the ceramic and other industries: the shaping of articles based on clay, whether by extrusion, pressing, plastic forming or slip-casting, requires a knowledge of these properties; in civil engineering, it is vital to be able to predict the behaviour of clay soils upon which buildings are placed. Other industries in which use is made of the flow properties of clay include mining, the petroleum industry, and paper and paint manufacture. In order to understand the complex nature of flow in clay–water and other suspensions, it is first necessary to study the flow of pure liquids.

DEFINITION OF VISCOSITY

Viscosity may be defined as the resistance of a liquid to flow. As a general rule, when a liquid flows, all parts of it must be in motion, so that successive planes of molecules slide over each other; thus, viscosity is a measure of the internal friction of a liquid. A quantity also used is *fluidity*, i.e. the ability to flow, defined as the reciprocal of viscosity.

Suppose that a liquid is flowing through a tube of uniform cross-section (Fig. 44). The layer of molecules in contact with the wall of the tube AB will be stationary and may be regarded as adsorbed on the solid surface; the adjacent layer of molecules has a small relative motion in the direction of flow indicated, the next a slightly greater velocity, and so on, with a maximum velocity at the centre of the tube, along the plane CD. The same

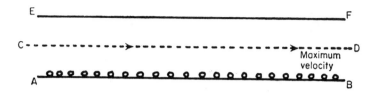

Fig. 44. Flow of liquid in a tube.

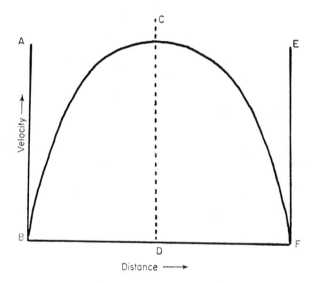

Fig. 45. Distribution of velocity.

velocity pattern in reverse is repeated from CD to the opposite wall of the tube, EF. If the actual velocities, relative to the walls of the tube, are plotted against distance from AB, the graph is a parabola with a maximum value at CD (Fig. 45). The velocity at any distance r from the wall of the tube is given by the equation:

$$v = k(R^2 - r^2)$$

where v = velocity, r = distance, and k and R are constants. These considerations are valid so long as the flow is *streamline*, i.e. all the planes of liquid move parallel to one another and to the wall of the tube. Above a certain critical velocity the flow becomes *turbulent* and the ordinary laws of viscous flow no longer apply. It is generally not difficult to maintain streamline flow when making laboratory measurements, and so turbulent flow need not be considered further.

The rate of variation of velocity with distance, dv/dr, is obviously important and is spoken of as the velocity gradient or *rate of shear*. This definition does not depend on the shape of the containing vessel but can be applied to any flow system. Consider now two adjacent planes of liquid, at a distance r apart (Fig. 46), each of cross-sectional area A, one moving relative to the other with a velocity v. *Newton's law of viscous flow* states

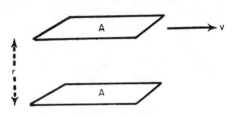

Fig. 46. Laminar flow.

that the frictional force F, acting between the planes, is proportional to the velocity gradient, and to the area A, so that:

$$F = \eta A \cdot dv/dr$$

or

$$F/A = \eta dv/dr$$

where η is a constant, called the *coefficient of viscosity*, or simply *viscosity*. Thus, F/A may be regarded as the force per unit area, or *stress*, required to maintain the flow of liquid. This law applies to any streamline flow for a pure liquid (i.e. a 'Newtonian liquid'), regardless of the shape of the containing vessel. It will be clear that the velocity gradient cannot generally be taken as v/r, since in many situations, as in considering flow through a tube (Fig. 44), dv/dr is not constant but varies with r.

MEASUREMENT OF VISCOSITY

The simplest method of measuring the viscosity of a Newtonian liquid is to determine the rate at which it flows through a tube when a constant pressure is applied. It can be shown that for a tube of uniform circular cross-section:

$$\eta = \frac{\pi P R^4}{8L.Q}$$

where η = viscosity, P = applied pressure, R = radius of cross-section, L = length of tube and Q = rate of flow (vol/sec). The above relationship is known as Poiseuille's equation, named after the originator of the method. One main disadvantage of the method is that the rate of shear cannot easily be controlled and moreover varies from point to point within the tube—a great disadvantage when dealing with clay and other suspensions, as will be seen later. As with all such viscosity measurements, care has to be taken to maintain a constant temperature, since η is markedly temperature-dependent for most pure liquids.

A rather more precise instrument is the *Couette* or rotating-cylinder viscometer. In this instrument, two concentric cylinders are immersed in the liquid. The outer cylinder is driven at a constant angular velocity by a synchronous electric motor; the inner cylinder is suspended from a light spring of beryllium–copper alloy, and carries a pointer which rotates with it. When the outer cylinder is in motion, the viscous drag of the liquid between the two cylinders exerts a couple or torque on the inner one, twisting the spring through an angle proportional to the torque. The angle of twist is recorded by the pointer on a circular scale, which may be calibrated directly in viscosity or stress units. If the angle of twist is θ, the spring constant k, radius of outer and inner cylinders respectively are R_2 and R_1, length of cylinders L, and velocity of rotation Ω (rad/sec), the stress is given by:

$$f = \frac{k.\theta}{2\pi R_1^2.L}$$

and the shear rate D is given by:

$$D = \frac{2\pi R_1^2.R_2^2 \Omega}{r^2(R_2^2 - R_1^2)}$$

where r is the distance of any plane in the liquid from the centre of suspension. From Newton's law, η is equal to f/D.

In practice, the difference between the radii of the cylinders, R_1 and R_2, is made as small as possible, so that the variation in r is kept small and the shear does not vary appreciably from point to point. If a complex system is being studied, it is necessary to measure f and D over a wide range of values and to plot f against D. The latter can be varied over a range by controlling the rate of rotation of the cylinders, Ω.

An even better control of shear rate is obtained with the *cone-and-plate viscometer*. In the latter instrument, a small sample of the liquid is placed on a flat plate which is rotated at a controlled but variable speed by a synchronous motor. Just touching the plate is a small-angle cone, suspended by a sensitive spring. Provided the cone angle is small (in practice less than $0.3°$) the rate of shear in the small volume between cone and plate is constant and is equal to $D = \Omega/\psi$, where ψ = cone angle and Ω = velocity of rotation of plate. Owing to the viscous drag of the liquid, the cone twists through an angle θ, so that the torque is equal to $k.\theta$. The stress f is then given by:

$$f = 3k.\theta/2\pi R^3$$

where k = spring constant and R = radius of cone. If desired, the viscosity, as before, can be calculated from the ratio f/D.

The *torsion viscometer*, used widely in the ceramic industries, has proved adequate for the control of casting slips but the readings obtained are on an arbitrary scale and cannot readily be converted to absolute units. It consists of a flywheel, carrying a pointer, suspended from a torsion wire of standard dimensions. Attached to the underside of the flywheel is a standard metal cylinder which can be immersed in the appropriate liquid or suspension. The latter, contained in a small beaker, is thoroughly stirred with an electric stirrer to break down *thixotropic structure* (see later). The sample is then placed under the flywheel without delay, so that the cylinder is completely immersed, and the flywheel rotated through 360° and clamped in position. The latter is then released and, depending on the viscosity of the suspension, rotates in the opposite direction due to the restoring force on the torsion wire. The amount of rotation beyond the first 360°, called the overswing, is read on a circular scale calibrated in degrees; should the flywheel fail to rotate by as much as 360°, the overswing is regarded as negative and read accordingly. The amount of overswing, expressed in degrees, is recorded as the fluidity of the suspension. After this reading has been taken, the flywheel is again rotated 360° and clamped. Exactly 1 min after taking the first reading, the flywheel is released and the overswing again recorded. If the suspension is *thixotropic*, the fluidity decreases and hence the overswing decreases. The amount by which the overswing is decreased is noted and is recorded as the 'thixotropy' of the suspension. (This quantity, like the fluidity, can occasionally be negative.)

NON-NEWTONIAN FLOW

Pure liquids such as water, alcohol, glycerine, etc., at constant temperature obey Newton's law of viscous flow. In abbreviated form, this may be written:

$$f = \eta D$$

where f = shear stress, D = shear rate and η = coefficient of viscosity. Thus, if a series of readings of stress for various shear rates are measured with a suitable viscometer, the graph of f against D for such liquids is a straight line (Fig. 47) the gradient of which is $1/\eta$. Suspensions in general do not obey Newton's law and give a curve of some kind when stress is

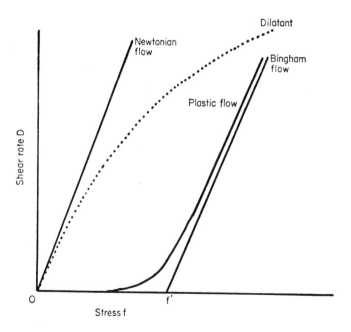

Fig. 47. The various types of flow curve.

plotted against shear rate, as shown. The curve may be either convex or concave to the stress axis and may or may not pass through the origin. In all such systems, the value of η is not constant but varies with shear rate.

Plastic flow

A plastic type of flow curve is given by the majority of clay–water suspensions and pastes in the flocculated state, shown in Fig. 47. An important characteristic of this curve is that it does not pass through the origin but intersects the stress axis at a point f'. This means that a definite minimum stress, called the *yield stress* or *yield value*, f', has to be applied before the system begins to flow. This definition of yield stress is the relevant one when dealing with viscometric measurements, but it is worthwhile noting that other definitions are sometimes employed and should not be confused with the present one. Whilst of small magnitude for suspensions, the yield value for clay–water pastes is very considerable and is an important factor in the shape retention of clay articles.

Bingham's law

Occasionally, a system gives a stress–shear graph that is linear but still shows a yield value (Fig. 47). In this instance the relationship between f and D may be written as:

$$f - f' = \eta_0 . D$$

where f' is the yield value and η_0 is a constant, sometimes called the *absolute viscosity*. Although few clay suspensions obey Bingham's law exactly, many approximate closely to it, only deviating at low shear rates. This behaviour is common for flocculated clay suspensions, particularly at high concentrations. The above equation may be written in an alternative form: if divided through by D, and the ratio f/D written, by definition, as η, the equation becomes:

$$\eta = \eta_0 + f'/D$$

From this equation it is clear that η, as normally defined, is greater than η_0 by a term f'/D, which varies with shear rate; moreover, it is also clear that as D becomes very great, this latter term becomes negligible and thus $\eta \to \eta_0$ as $D \to \infty$. In order to distinguish it from the absolute viscosity, η is frequently called the apparent viscosity. Plastic flow has also been described by equations of type:

$$f^n = k.D$$

where n and k are constants, but these are on the whole less satisfactory.

Dilatancy

A limited number of systems exhibit *dilatant* behaviour, giving a flow curve shown by the dotted line in Fig. 47. This type of behaviour is shown mainly by non-plastic materials such as alumina, flint, quartz, etc. Purely dilatant systems have no yield value and, as is evident from the curve, the viscosity of such systems, defined by f/D, *increases* with increasing shear rate.

The reason for dilatant behaviour is not fully understood, but it may be due to the 'squeezing out' of water from between the particles of the suspension at the places of greatest shear; this is consistent with the observation that dilatancy is most marked with suspensions of coarse particles, which have little inter-particle cohesion. According to Reynolds, however, dilatant systems are close-packed when at rest, but on being sheared rearrange to form a loosely packed structure containing large voids. There is insufficient water to fill these voids and so the viscosity increases markedly. Dilatancy is frequently observed in deflocculated clay systems at high rates of shear. In these systems the flow curve is often S-shaped, indicating plastic flow at low shear rates, giving way to dilatancy at higher shear rates.

Pseudo-plasticity

This is the term given to systems that give a flow curve similar in shape to plastic systems but having no yield value, i.e. the curve passes through the origin. It is said to occur with suspensions of asymmetric particles that tend to orientate themselves along the direction of shear. It may in practice be difficult to differentiate pseudo-plasticity from true plasticity.

Mechanism of plastic flow

As previously explained, the apparent viscosity of clay suspensions decreases with increasing shear rate. This behaviour, and the existence of a yield value, is generally believed to be due to the formation of an internal structure in the suspensions. Flocculated systems, as previously explained, have a strong tendency to form flocs or aggregates. In a suspension at rest therefore, a three-dimensional network of particles is formed, which must first be broken down before flow can occur; thus, the system possesses a yield value. Breakdown is not complete, however, at the commencement of shear; large aggregates still remain although the continuous structure has been destroyed. The effect of increasing the shear rate is to break down these aggregates progressively into their individual particles, thus reducing their mutual interference and thereby reducing the viscosity.

However, there is a natural tendency for the clay particles to re-aggregate, and for any given shear rate there is an equilibrium between this tendency to re-form the structure and the breakdown caused by shearing. On removing the shear, therefore, the structure builds up again to its initial state.

With asymmetric particles like those of clay, a secondary effect of aggregate breakdown is the progressive alignment of the individual particles in the direction of shear, thus reducing further the resistance to flow. This progressive breakdown with increasing shear is probably the reason for the curvature in the f–D graph.

Thixotropy

Many clay suspensions, if allowed to stand undisturbed for some time, are observed to thicken up, i.e. become more viscous; in extreme cases (e.g. clays treated with excess electrolyte) the vessel containing them may be inverted without causing the suspension to run out. On vigorous stirring, such suspensions become quite fluid again, reverting to their original condition when stirring ceases, and so on. This reversible, time-dependent property is known as *thixotropy*. As might be expected, thixotropic slips show a reduction of viscosity with increasing shear rate; their flow curves usually show a yield value, with marked curvature in the region of lowest shear rate.

Sedimentary kaolinites treated with sodium carbonate in slight excess exhibit marked thixotropy; montmorillonites may have such a high degree of thixotropy that they form *gels* when undisturbed. It should be noted that true thixotropy always implies a measure of time-dependence; systems that show a decrease of viscosity with increasing shear rate only are not necessarily thixotropic.

Cause of thixotropy

It has been observed that in thixotropic suspensions that have been at rest, the Brownian motion is suppressed, suggesting that some kind of internal structure is built up. For clays, so-called 'house of cards' or 'scaffold' structures have been envisaged, in which the platy particles of clay are linked together in a three-dimensional network extending throughout the suspension. Particle linkages have also been associated with a yield value, of course, but the essential distinction is that thixotropic structures require a measurable time for breakdown under shear and for natural build-up.

Thixotropy is most marked in imperfectly deflocculated kaolinite suspensions, although it should be remembered that a structure capable of extremely rapid build-up or breakdown might not appear thixotropic when tested in an ordinary viscometer. The customary procedure when using a viscometer of the rotating cylinder or cone-and-plate type is to note the stress reading obtained at the smallest practicable shear rate. This reading is in general not a steady one but for a thixotropic system gradually decreases, and a very long time is normally necessary to obtain a steady reading. For this reason, it is customary to allow an arbitrary time of 10 sec after the first application of shear, before the stress reading is taken. The shear rate is then increased a further step, and again 10 sec allowed to elapse before taking a stress reading. This process is continued up to the maximum value practicable, after which the process is repeated in the reverse direction, by decreasing the shear step-wise. In this way, a true thixotropic system gives a curve similar to that of Fig. 48, in which the upward and downward curves do not coincide but form an anticlockwise loop.

This hysteresis effect can be explained as follows. On first applying a shear, the thixotropic structure begins to break down, but this effect is not instantaneous and so the reading of stress is greater than the 'equilibrium stress'. Thus, the 'up-curve' is always to the right of the hypothetical 'equilibrium curve'. On decreasing the stress, the opposite effect occurs; the structure attempts to build up and so the 10 sec reading is always less than the equilibrium value, i.e. the down-curve is on the left of the 'equilibrium curve'.

Moore and Davies (1959) derived an equation relating the breakdown and build-up of thixotropic structure to time and showed that experimental data agreed reasonably well with their equation. They considered thixotropic build-up complete when the maximum number of particle–particle links was present; the degree of build-up of any system was defined as the number of links λ existing at any instant, expressed as a fraction of the maximum. Thus, for a rested system, $\lambda = 1$, indicating complete build-up, whilst for complete breakdown under shear, $\lambda = 0$. If λ_0 denotes the value of λ at zero time, and λ_D the equilibrium value of λ, then after any time t the value of λ is given by the equation:

$$\lambda - \lambda_D = (\lambda_0 - \lambda_D)\exp-(a+bD)t$$

where a and b are constants of build-up and breakdown. It is clear from the equation that the decay of thixotropic structure with time follows an

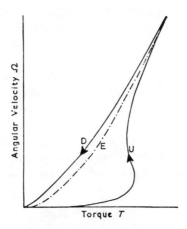

Fig. 48. Thixotropic loop.

exponential law and that theoretically an infinite time is required to reach equilibrium.

The width of the hysteresis loop obtained by the standard procedure of 10 sec readings has been considered to be a measure of the degree of thixotropy; it actually represents the maximum stress deviations from equilibrium over a 10 sec period on each side, at an arbitrary shear rate. More precise information would be needed to describe thixotropy fully, however; it would be necessary to know the maximum strength of the fully built-up structure, the rate of breakdown under shear, the dependence of breakdown on shear and the rate of natural build-up.

Significance of 'bottom curvature'

If the linear part of the plastic flow curve of Fig. 47 is produced to meet the stress axis, the value of the intercept is sometimes called the 'virtual yield value', being the yield value that would be obtained if the system obeyed Bingham's law down to the lowest shear rates. The difference between this 'virtual yield value' and the true yield value is clearly a measure of the amount of 'bottom curvature' or the extent of the deviation from Bingham's law. This deviation may be due to the progressive breakdown of thixotropic structure as the shear rate is increased, resulting in a gradual transition from a highly structured system, with high viscosity, at low shear rates, to a system of little or no structure and low viscosity, at high shear rates.

The author has found that the flow curves of flocculated, non-dilatant clay–water suspensions can be represented fairly accurately by the following equation:

$$f - f' = \eta_0 D + gbD/(aD + b)$$

where f = shear stress, f' = yield value, D = shear rate, η_0 = absolute viscosity, a and b are coefficients of breakdown and build-up respectively, and g is another constant related to the degree of flocculation.

The expression is essentially the Bingham equation with an extra term that describes the bottom curvature, the magnitude of which is given by gb/a. This modified equation is a hyperbola, one asymptote of which is the 'Bingham line':

$$f - f' = \eta_0 D + gb/a$$

Although the yield value f' and the 'bottom curvature' gb/a are both high for flocculated systems, they are not necessarily related, since the factor $b/(aD + b)$ in the equation represents the state of build-up of thixotropic structure at the shear rate D and thus has an element of time-dependence. It is just possible that f' is a static yield value, whilst gb/a is a shear term derived from particle collisions and cohesions in a system undergoing shear. The validity of the equation for non-aqueous systems has not been fully investigated, but it seems that for a finite value of f', a polar liquid is required, but appreciable thixotropy and 'bottom curvature' can be present for clays suspended in non-polar liquids.

Rheopexy

This term was originally defined as the gradual solidification of thixotropic systems by gentle agitation and there is no doubt that some thixotropic clays do behave in precisely this manner. It is generally believed that *gentle* agitation assists the clay particles to occupy the most favourable positions for formation of a gel but of course more violent shearing would inevitably destroy any such structure.

However, since about 1960, the term has broadened to include all systems that show an increase of viscosity with time when subjected to shear, i.e. shear-thickening or the reverse of thixotropy. By this definition, the shear stress–shear rate curves of such a system have hysteresis loops that are *clockwise*. As will be seen later, however, clockwise hysteresis loops may not always indicate rheopectic behaviour.

The effect of solid concentration on viscosity

The apparent viscosity of a suspension is in general higher than that of the suspending medium; many investigations have been concerned with the relationship of apparent viscosity to concentration, but unfortunately few have involved independent determinations of yield value, or absolute viscosity. Since for non-Newtonian systems the apparent viscosity also varies with shear rate, it is necessary to choose an arbitrary shear rate at which all measurements are to be made.

There are various mechanisms by which suspended particles influence the viscosity of the suspension; at low concentrations, local disturbance of the flow pattern of the liquid occurs, but at higher concentrations overlap of the centres of disturbance develops, and at still higher concentrations collisions between particles have to be taken into account. For charged particles, the forces required to shear off the double layers may need to be considered—the so-called *electro-viscous effect*. Again, many investigations concerned only spherical particles, since the latter are simpler to treat mathematically; various corrections have been proposed for ellipsoidal, rod-shaped and tabular particles. Much of the work published concerns non-clay materials, such as polymers, textile fibres, glass spheres, etc., but some of the equations may be adapted for clays.

Considering only the first effect mentioned, namely, the disturbance of the flow pattern of the liquid medium, Einstein put forward the following equation for dilute (less than 2% by volume) suspensions of uncharged, spherical particles:

$$\eta_s = \eta_m(1 + kc)$$

where η_s = viscosity of suspension, η_m = viscosity of pure liquid, c = concentration and k = a constant, the value of which is about 2·5 for spheres. As is customary with rheological work, the concentration is measured as (volume of solid) ÷ (total volume of suspension) and is thus always less than unity. It is occasionally expressed as a percentage. The value of k is found to be sensitive to particle shape and is greater than 2·5 for platy, rod-shaped and needle-shaped particles, etc., but less than 2·5 for deformable particles such as emulsions. With considerable departure from sphericity, however, the shear dependence becomes pronounced and more complicated equations have been developed, involving a shape factor.

For much more concentrated suspensions, of up to 50% or so, which are common in industrial casting-slips, the particle–particle collisions become the dominant factor in determining viscosity, and since this outweighs completely the viscosity of the liquid, the system as a whole is not appreciably sensitive to changes of temperature. Equations for these concentrated suspensions are often based on a power law, a typical example being that developed by Norton and his associates:

$$\eta_s = \eta_m(1-c) + k_1 c + k_2 c^n$$

where η_s = viscosity of suspension, η_m = viscosity of liquid, c = concentration, k_1 and k_2 are constants depending on the particle size, and n is another constant, the value of which is approximately 3 for flocculated suspensions. The first two terms correspond to Einstein's equation with a correction for the reduction in the effective volume of the liquid by the presence of the solid particles. The third term is the 'interference factor' associated with particle collisions. For deflocculated suspensions, the value of n was found to be approximately equal to 12, with very small values for k_1 and k_2.

Norton et al. (1944) also found empirical relationships between yield value and concentration, of the form:

$$f' = k_3 c + k_4 c^3$$

where f' = yield value and k_3 and k_4 are constants. These equations were claimed to be valid for concentrations up to 20%, but beyond this some discrepancy can be expected.

For still higher concentrations, an equation of the type used by Mooney may be more applicable:

$$\eta_s/\eta_m = \exp[ac/(1-kc)]$$

where a is a constant and the other symbols are as previously. The author found that if η_s is measured as the *absolute* (Bingham) viscosity, the equation is obeyed reasonably well by clay suspensions up to 50% by volume.

The effect of ion exchange on viscosity

It will be clear from the previous chapter that the forces between suspended particles, on which viscosity depends, are in turn dependent on the nature of the cation associated with the double layer. Thus, flocculating cations such as OH_3^+, Ca^{2+} and Mg^{2+} favour the formation of particle–particle links, leading to high yield values and thixotropic structures; the viscosity of such systems is correspondingly high. If the above cations are replaced by Na^+, K^+, Li^+ or NH_4^+, which favour a high zeta-potential, the interparticle forces of attraction are outweighed by the electrostatic forces of repulsion and a deflocculated system results. Such a system therefore has a small or negligible yield value and a low viscosity.

Since clay suspensions obey Bingham's law approximately, it is convenient to express the effect of deflocculants in terms of the equation constants. Recent work has shown that the yield value f' diminishes rapidly as a suspension is deflocculated, becoming very small or zero near the optimum. The absolute viscosity, η_0, is apparently little affected by deflocculation, as might be expected, since it is in effect the limiting viscosity at the point where all internal structure is broken down by shear.

Using the alternative form of the Bingham equation:

$$\eta = \eta_0 + f'/D$$

it is clear that if the law is obeyed, the graph of η against $1/D$ is a straight line of gradient f' and intercept η_0. When this is done for a number of clay–water suspensions of the *same* concentration but in different states of deflocculation, as shown in Fig. 49, a series of lines is obtained, all of differing slope but tending towards the same intercept. It will be clear that as the number of milliequivalents of sodium is increased, the system becomes progressively more deflocculated until the optimum amount is reached, when the slope f' is zero or nearly so. Over the range of shear rate covered by Fig. 49, good straight lines are usually obtained, but at *lower* shear rates (i.e. further from the origin) deviation occurs, the graph bending towards the shear axis.

Considering the deviation from Bingham behaviour in more detail, the effect of deflocculation may be illustrated by reference to the equation:

$$f - f' = \eta_0 D + gb . D/(aD + b)$$

As before, f', the yield value, decreases progressively as deflocculation proceeds, whilst η_0 remains unaltered. At the same time, the overall

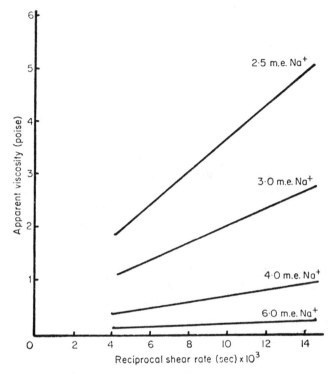

Fig. 49. Deflocculation and Bingham constants.

magnitude of the last term decreases, so that the amount of 'bottom curvature' is also reduced. Apparently, the value of g is decreased by deflocculation, but a and b are also affected in a way that is not fully understood. It seems likely, however, that g, b and a all tend to zero as deflocculation becomes complete.

Application to slip-casting

In the control of industrial casting-slips, it is important to control, among other things, the solid concentration of the slip. This is usually done by weighing an accurately measured volume, expressing the result simply as the overall slip density in $g\,ml^{-1}$ or ounces per pint. For additions of deflocculant, based on the weight of solid material, it is then necessary to calculate from this the amount of solid per unit volume of slip. This can readily be done by the use of Brongniart's formula:

$$w = 1000(S-1)\frac{d}{d-1}$$

where w = weight of solid material in $g\,l^{-1}$ of slip, S = slip density in $g\,ml^{-1}$ and d = average density of solid material in $g\,ml^{-1}$. It was originally formulated in British units as:

$$w = (P-20)\frac{d}{d-1}$$

where w is in ounces per pint, and P (the slip density) is also in ounces per pint. These very useful formulae can be derived from first principles on the assumption that the volumes of water and solid are strictly additive, which is sufficient for all practical purposes.

Table 20
Adjustment of Flow Properties

Addition	Fluidity	Thixotropy
Water	Increase	Small decrease
Deflocculant	Increase	Decrease
Clay	Decrease	Increase

Apart from slip density, which is of the order of $1\cdot6$–$2\cdot0\,g\,ml^{-1}$ for clay-based slips, two other parameters are required for adequate control of the quality and rate of casting. These are the fluidity and the thixotropy, as measured on the industrial torsion viscometer described earlier. Normally the ratio of sodium silicate to sodium carbonate used in the deflocculant mixture is maintained constant, but the total deflocculant and the slip density may be adjusted marginally to give the desired result. Typically, a casting-slip when properly adjusted could have a fluidity of, say, 345° and a thixotropy of 25°. Where either of these values is too high or too low, appropriate additions of clay, water or deflocculant are made. Generally, it is not possible to adjust fluidity without automatically changing the thixotropy in the opposite direction, but these opposing changes are not necessarily equal and it is therefore possible to obtain the required adjustment by the judicious addition of two or more components. These effects are summarised in Table 20.

The various casting faults caused by incorrect adjustment of fluidity and thixotropy depend to some extent on the type of casting-slip, but typical examples of such faults and their probable causes are shown in Table 21.

Table 21
Casting Faults for Sanitary Earthenware Slips

Flow characteristic	Casting-fault
Fluidity too high	Slow casting rate Cracking 'Wreathing'
Fluidity too low	Pin-holing Bad draining
Thixotropy too high	'Flabby' casts Bad draining Slow drying
Thixotropy too low	Brittle casts Difficulty in 'fettling' Slow casting rate 'Wreathing' Cracking

'Ageing' of clay suspensions

Up to the present, it has been assumed that all flow measurements are made on stable suspensions, i.e. those whose flow properties do not change appreciably on storage. It is well known, however, that a freshly made clay suspension undergoes marked changes in viscosity during storage; in general, the viscosity increases continuously and irreversibly with time, these changes continuing for a period of at least several months. This effect is usually referred to as 'ageing'. Plastic masses of clay also exhibit ageing and for this reason it has been common practice in industry to store plastic clay for one or two months before use to achieve stabilisation. It is obviously an advantage to reduce ageing time as much as possible, for example by shearing or stirring clay suspensions and in general by putting 'work' into the clay. Homogenising, ultrasonic treatment and ball milling all reduce ageing time, but the time required for stabilisation is still considerable. Worrall and Basu (1965) have found that further changes could still be detected in a slip that had been ultrasonically treated for as much as 200 hr.

Since the process of doing work on a suspension changes its flow properties, it is evident that the shear produced by a viscometer in making these measurements will have the same effect. This is illustrated in Fig. 50, which shows the result of repeated up-and-down measurements or cycles on a freshly made, flocculated clay suspension. In cycle 1, the down-curve falls to the right of the up-curve, showing that the stress and therefore the

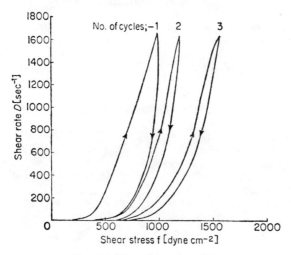

Fig. 50. The effect of ageing.

viscosity have been increased by the initial shear, resulting in a *clockwise* loop. Cycles 2 and 3 also consist of clockwise loops, each cycle falling to the right of the preceding one. It has been claimed that by repeated cycling, stable curves that do not drift to the right can be obtained, but this method of stabilisation is very tedious. It should also be noted that this shear-accelerated ageing may completely mask the natural thixotropy of slip and, indeed, the clockwise loops have sometimes been taken to indicate 'negative thixotropy'. The latter term is misleading because the phenomenon in question is irreversible and has no connection with thixotropy, which is time-dependent.

Work by the author indicates that these irreversible changes associated with ageing are caused by a progressive and irreversible breakdown of aggregated clay particles by the action of water, in some cases assisted and accelerated by shearing. This has been confirmed by size distribution measurements, which show a progressive increase in the proportion of fine particles as the system is aged. Evidently, the breakdown of aggregates to ultimate particles increases the *available* surface area, so that the contact area between particles is increased, which in turn increases yield value and viscosity. Deflocculated systems remain deflocculated after ageing and so the viscosity changes are much less than with flocculated systems.

REFERENCES

MOORE, F., and DAVIES, L. J., *Trans. Brit. Ceram.'Soc.*, **58** (1959) 470.
NORTON, F. H., JOHNSON, A. L., and LAWRENCE, W. G., *J. Am. Ceram. Soc.*, **27** (1944) 149.
WORRALL, W. E., and BASU, M. K., *Trans. Brit. Ceram. Soc.*, **64** (1965) 61.

READING LIST

E. C. BINGHAM, *Fluidity and Plasticity*, McGraw-Hill, 1922.
F. MOORE, *Rheology of Ceramic Systems*, Institute of Ceramics Textbook Series, Elsevier Applied Science Publishers, 1965.
F. MOORE, and A. DINSDALE, *Viscosity and its Measurement*, Institute of Physics, 1962.
G. W. SCOTT-BLAIR, *Elementary Rheology*, Academic Press, 1969.
D. J. SHAW, *Introduction to Colloid and Surface Chemistry*, Butterworth, 1970.

Chapter 7

The Plasticity of Clays

DEFINITION OF PLASTICITY

It is a characteristic of clay that when mixed homogeneously with water in suitable proportions it becomes a coherent mass, capable of being moulded under stress to any desired shape, and capable of retaining that shape when the moulding stress is removed. These properties constitute a definition, so far as one exists, of *plasticity*. From the rheological point of view, plastic masses of clay are not unlike suspensions, the main distinction being that the solid concentration is higher in the former, resulting in a higher yield value and the property of shape retention.

PLASTIC FLOW

It is immaterial for the purpose of definition what form the plastic body takes, but perhaps the most convenient way of illustrating plastic flow is to enclose the clay in a tube of constant cross-section, apply a pressure at one end, and observe the rate of flow of clay through the tube. If rate of flow is plotted against pressure, a curve similar to the plastic flow curve of Fig. 47 is obtained.

Absolute values of stress and shear rate cannot of course be obtained by this procedure, and according to Buckingham and Reiner, even a true Bingham body under these experimental conditions yields a curve with a slight 'bottom bend'. However, a bottom bend is obtained for clay suspensions even with a cone-and-plate viscometer, proving that they do deviate from the Bingham law; moreover, the amount of bottom curvature obtained can in no case be fully accounted for on the Buckingham and Reiner theory. The flow curves for plastic masses are thus very similar to

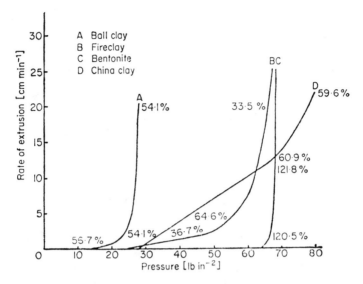

Fig. 51. Flow curves of various clays.

those obtained for suspensions, except that the yield values are much higher.

If the pressure–flow curves of different clays are compared, it is apparent that there are striking differences, as shown in Fig. 51. Although it is impossible in practice to work with the same water content for all the different clays, several important facts emerge from the curves shown. First, the montmorillonite has the smallest amount of 'bottom curvature' and approaches most nearly to Bingham behaviour; second, it has the highest yield value, despite the high water content. Third, china clay shows the greatest deviation from Bingham behaviour, whilst ball clay and fireclay are intermediate in general characteristics. Since the relative plasticities are usually considered to be in the order:

montmorillonite > ball clay > fireclay > china clay

it appears that high plasticity is characterised by a high yield value and a small degree of 'bottom curvature'. The author has found that the flow curves of plastic clays in general obey the same empirical equation as suspensions:

$$f - f' = \eta_0 D + gb D/(aD + b)$$

Since the magnitude of the 'bottom curvature' is given by gb/a and the yield value is f', the ratio $f'/(gb/a)$ is an indication of the degree of plasticity. It has been found that this ratio is largely independent of water content and increases in the recognised order shown above for the different clays. The same ratio has also been found to decrease linearly with incremental additions of sand, the plasticity of the latter, for all practical purposes, being zero. Typical values of this ratio for various clays are given in Table 22.

Table 22
Values of the Ratio $f'/(gb/a)$ for Various Clays

Clay	$f'/(gb/a)$
Ball clay (natural)	2·80
Na-ball clay	5·60
H-ball clay	3·33
Ca-ball clay	2·33
China clay	0·54
Fireclay	1·23
Montmorillonite	4·70

Apart from pressure–flow curves, it is difficult to measure the rheological properties of plastic clay by continuous flow methods, as in a viscometer, since there is always a degree of slippage between the cylinders or cone and plate, and the clay. Although toothed wheels have been used to overcome slippage, fracture of the clay can still occur after a short time and so the method is not completely satisfactory. For this reason, much of the research done on plastic clay has concentrated on treating it as a solid on which measurements of stress and strain can be made. Thus, one can measure the extension or compression on a clay cylinder produced by a stress, or the amount of twist in a cylinder when a torque is applied.

STRESS–STRAIN MEASUREMENTS

At first sight, stress–strain measurement may seem the more obvious and direct way of studying plastic behaviour, but it suffers from many drawbacks and is not always as informative as might be expected. If a bar of an *elastic* substance, preferably a metal, is stretched by applying a stress, it

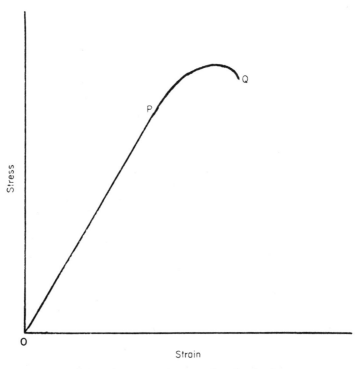

Fig. 52. Stress–strain curve for plastic clay.

is found that over a certain range the strain is proportional to the stress, i.e. *Hooke's law* is obeyed. The strain is usually expressed as a fraction of the original length; thus, if the amount of elongation of a bar of length l is δl, the *strain* is $\delta l/l$. If the stress f is plotted against the strain, as in Fig. 52, a straight line OP is obtained. In other words,

$$f = E \cdot \delta l/l$$

where E is a constant called the modulus of elasticity or Young's modulus. If further stress is applied, a point is reached (marked P in the diagram) where the bar begins to *yield* or *flow*; from this point onwards, Hooke's law is no longer obeyed and the graph is curved, as shown, from P to Q. Finally, if sufficient stress is applied, the bar breaks and the stress falls off rapidly, as shown. The portion of the curve PQ is the *ductile stage*, P is the *proportional* or *elastic* limit and Q is the *fail point*. A peculiarity of metals is that, having stretched a bar beyond the elastic limit without fracture, the

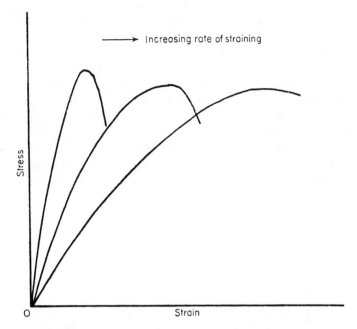

Fig. 53. Effect of loading rate on stress–strain curve.

experiment can be repeated, producing another straight line with a larger gradient, indicating that the metal is now harder; this property is known as *strain-hardening* or *shear-hardening*.

Many of the concepts pertaining to metals have been utilised in studying clays, although sometimes they can be misleading. In a series of experiments on clay bars, Macey found that under tensile stresses, Hooke's law was not obeyed, even over a restricted range, so that the graphs of stress f against extension E were curves. Moreover, the results depended very largely on the rate at which the stress was applied, different rates of loading giving different curves, as shown in Fig. 53. As the rate of loading increases, the clay effectively becomes less and less brittle, in contrast to many other substances. Macey deduced the following relationship between stress and elongation:

$$f = B(1 - \exp - aE)$$

where f is the stress, E the elongation, and B and a are constants. However, there did not appear to be any clear connection between the values of these constants and plasticity.

Whittaker measured the torque required to twist a clay bar through a measured angle, for a series of moisture contents. These results did seem to show that Hooke's law was obeyed over a limited range, and Whittaker defined a 'yield value' Y as the stress at which Hooke's law just ceased to be obeyed, i.e. the point at which the stress–strain graph just became curved. From this point up to the fail point was defined as the amount of *extensibility* E, analogous to the ductile region for a metal. Whittaker then defined a plasticity index as (yield value) × (extensibility) or $P = Y \times E$. It was claimed that this quantity varied only slightly with water content and was closely related to particle size. The most difficult measurement in this method would seem to be the yield value, since it is hard to observe just

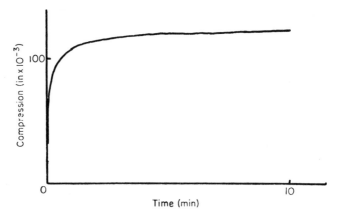

Fig. 54. Deformation–time curve of plastic clay.

where the graph ceases to be linear. Moreover, this definition of yield value departs from that generally accepted by rheologists.

A phenomenon often quoted in all stress–strain measurements on plastic clay is that of *shear-hardening*. If a load is applied to a bar of plastic clay, deformation takes place rapidly at first but gradually slows down until it is scarcely perceptible (Fig. 54). This is often described as shear-hardening, but it is clearly different from that associated with metals. Norton found similar results with glasses and claimed that two types of flow occur in these circumstances: (a) elastic flow, which is partly reversible and accounts for the initial rapid deformation; and (b) viscous or plastic flow, which proceeds at a constant rate. The elastic part of the flow ceases when restoring forces that are built up equal the applied stress. This theory may possibly

The Plasticity of Clays

account for the type of result shown and is borne out by the fact that a limited amount of reversible elasticity does occur with plastic clays. The 'shear-hardening' referred to, combined with the yield value, may account for the so-called 'memory' of clays.

In some recent experiments by Astbury, clay cylinders were subjected to repeated cycles of torsion in the forward and reverse directions (Fig. 55). If a torque is applied, at a regulated rate, to a clay cylinder, and the angle of twist is plotted against the torque, a curve OA is obtained, as shown. Having applied a maximum stress at A, the latter is then reduced, when it is found that the return path is not AO but AB, showing that there is little recovery and that much of the original deformation remains, as would be

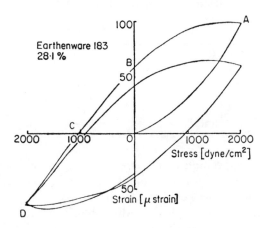

Fig. 55. Cyclic stress–strain curve.

expected of a plastic body. On reversing the stress, the original strain is gradually removed, but does not completely disappear until the point C is reached. On further increasing the reverse stress to a maximum value at D, the strain is then built up in the reverse direction, as shown by the curve CD. Finally, on reducing the negative stress, partial recovery occurs, and the curve traces out a second cycle, which does not necessarily coincide with the first. Repeated cycling does eventually give a reproducible cyclic loop, however, in a manner analogous to repeated cycling of a suspension in a viscometer.

It was suggested by Astbury that an ideal plastic body would show zero recovery and theoretically would follow a rectangular path enclosing the loop. The ratio of the area of the loop to that of the rectangle was thought to give a measure of the plasticity of the clay. For china clay the ratio was 0·5 and for montmorillonite 0·33, so that on this basis the china clay is the more plastic. It should be realised, however, that this index is essentially a measure of non-recovery, whereas the practical notion of plasticity is almost certainly based on other criteria such as water tolerance, binding power, etc. A mathematical analysis of Astbury's results enabled certain parameters to be calculated, related to the energy stored in the clay. More recently, Moore and Astbury have developed a compression test, in which

Table 23
Plasticity Index for Various Clays Obtained by the Compression Plastometer

Clay	Plasticity index
China clay	0·63
China body	0·41
Ball clay	1·40
Pottery body	0·44–0·50
Fireclay	0·80
Montmorillonite	1·1–1·3

a cylinder of clay is compressed progressively under increasing loads. The ratio of the load stress at 10% compression to that at 50% compression is used as an index of plasticity. Briefly, the stress at 10% compression is taken as an indication of yield value, which should be high for a plastic clay; the stress at 50% compression is considered to be an indication of the resistance to deformation at high stresses, which of course should be low for a plastic clay. It is frequently found that the stress–compression curves for very plastic clays 'turn over' at or about 50% compression, thus giving values of the index greater than 1. Some typical values of the plasticity index for various clays, obtained by this method, are given in Table 23.

THE ATTERBERG PLASTICITY INDEX

Generally speaking, the more plastic a clay, the more water it will tolerate without becoming fluid; in other words, the range of water content over which it is plastic is very wide. This is the basis of the Atterberg plasticity index, defined as follows. If water is added, little by little, to a dry clay, a stage is first reached where the clay just becomes plastic and coheres; this corresponds to the minimum amount of water required to form a stable film around each clay particle and is expressed as a percentage of the dry weight of clay. As more water is added, the plastic clay becomes softer and eventually reaches a stage where it begins to flow under its own weight; this corresponds to the water content required to reduce the yield value below an arbitrary level and is therefore less well defined than the first stage, but methods have been devised for its measurement that give reproducible results. The first stage is called the *plastic limit*, the water content being denoted by P_w; the second stage is the *liquid limit*, at which the water content is denoted by L_w. The Atterberg plasticity index is then given by $L_w - P_w$. Although it is unrelated directly to the mechanical behaviour of clay, the Atterberg index does give values that are in accord with practical experience.

There are many factors that govern the Atterberg limits for clay. At the plastic limit, there is just sufficient water to form stable films and so the magnitude of P_w will depend on the specific surface area of the clay and to some extent on the nature and amount of the exchangeable cations. At the liquid limit, the water films have built up to their maximum thickness and therefore much of the water present is 'free' or bulk water. Calculations of the maximum film thickness for flocculated kaolinites up to the liquid limit have given improbably high values of the order of 1000 Å, and it now seems likely that much of this water is not held as films but is mechanically entrapped by a network of clay particles forming the so-called 'house of cards' or scaffold structure. The amount of water that can be entrapped in this way obviously depends on the ability of the clay particles to form links with one another, that is, on the degree of flocculation. Hence, Ca-kaolinites have higher liquid limits than Na-kaolinites. In addition, the entrapment of water will also depend on the particle size; the smaller the clay particles, the greater the efficiency of the entrapping network.

For montmorillonites, the factors are somewhat different in their effect. The factors affecting the *plastic limit* apply in a similar way to montmorillonites; indeed, the effect is enhanced because of the greater number of exchangeable cations present. On the other hand, the entrapping mechanism is no longer so important in governing the *liquid limit*, since the lattice of montmorillonite is more or less penetrable by water in all circumstances. In general, Na- and K-montmorillonites have higher liquid limits than the corresponding Ca and Mg clays; this is possibly because the high zeta-potential associated with Na, K and Li ions renders the lattice more easily penetrable and furthermore the alkali ions may be able to exert a stronger orientating effect on the water molecules than do Ca and Mg, thus increasing the thickness of the internal water films. Some representative values for P_w and L_w are shown in Table 24.

The strong influence of the exchangeable cations on the liquid limit is readily appreciated from the values quoted. Accordingly, the plasticity index, $L_w - P_w$, is greater for, say, Na-montmorillonite than for Ca-montmorillonite, but whether this accords with common experience is not absolutely clear. The index for Ca-kaolinite is somewhat greater than that for Na-kaolinite, this result being more or less in agreement with the generally accepted view.

A method for the determination of the Atterberg indices that has been used in the ceramic industry is as follows. A suspension of the clay is made

Table 24
Plastic and Liquid Limits for Various Kaolinites and Montmorillonites

Clay	P_w	L_w
Ca-kaolinite	36	73
Na-kaolinite	26	52
Ca-montmorillonite	65	166
Na-montmorillonite	93	344

by stirring with excess of water to give a homogeneous slip of creamy consistency. The latter is poured on to a porous surface (e.g. a slab of plaster of Paris) and after allowing a minute or so for dewatering, a knife is dipped into the slip at intervals, noting whether the incision made disappears. When the stage is reached at which the incision just remains, a sample of the clay from around the same area is removed, weighed and dried, and its moisture content determined. This represents the stage at which the transition from a suspension to a paste occurs and therefore corresponds to the liquid limit. The clay is then allowed to dry further on the porous slab until it can be peeled off. A portion of the plastic clay is then kneaded with the hands to assist further evaporation of water, until, on rolling between the fingers, it commences to crumble, and loses cohesion. This stage corresponds to the plastic limit. The moisture content of the clay at this stage is determined by drying, to obtain the value of P_w. It should be pointed out that whereas for kaolinitic clays the drying can be carried out adequately at 110°C, montmorillonite presents some difficulty, since inter-lattice water is lost over a range of temperature from 100°C to about 350°C; indeed, in some instances it may be difficult to define just where all the inter-lattice water is lost and the dehydroxylation of the lattice begins. A more prolonged period of drying may be necessary, therefore, or a higher drying temperature may be employed.

Another test for plasticity based on water tolerance is that of Pfefferkorn. A series of cylindrical test-pieces of clay is prepared, of standard dimensions, covering a range of moisture contents. Each piece in turn is subjected to a standard compression obtained by allowing a standard disc and guide rod to fall on it from a fixed height. The deformation produced in each test-piece is calculated as the ratio of original length to deformed length. The deformation ratio is plotted against the moisture content, to give a smooth curve. The Pfefferkorn index is the moisture content corresponding to a deformation ratio of 3:1. This test has also found considerable application in industry and the results correlate approximately with accepted values.

THEORY OF PLASTICITY

A plastic mass of clay behaves rheologically very like a suspension, as has been shown; therefore, the forces between the particles are governed by similar factors, the main difference being that the particles in a plastic mass are closer together. In the present discussion it will be assumed that the liquid medium or plasticiser is water, but it should be remembered that other polar liquids, e.g. glycerol and ethylene glycol, can also act as plasticisers.

The clay particles in a plastic mass carry the normal negative charges and surrounding them is a film of adsorbed water molecules, bonded by electrostatic forces. Thus, the water present is of two kinds: (a) 'bound' water, present in the films; and (b) 'free' water, in excess of that required to form films. The thickness of the water films, as with suspensions, depends on the surface density of charge and on the exchangeable cations present. The available surface area of the particles is also important, since the amount of 'bound' water is greater, the greater the surface area on which it can be adsorbed. Hence, plasticity is associated with very small particles, i.e. those of colloidal dimensions. Another important factor is the shape of the particles. For a given mass, a thin platy or fibrous particle possesses a greater surface area than spheres or cubes; therefore, the platy particles of clays are conducive to a high specific surface, which in turn favours plasticity.

It is generally accepted that the particles of clay in a plastic mass are in positions of equilibrium, the repulsive forces between them balancing the forces of attraction. The former are electrostatic in origin, associated with the zeta-potential, whilst the latter derive partly from van der Waals forces at short range and partly from surface tension due to the water. The water films probably act as a lubricant, facilitating the sliding of particles over one another when the mass is sheared, whilst cohesion is ensured by van der Waals forces and surface tension. Where the van der Waals forces outweigh the repulsion forces, as in flocculated systems, an internal structure or network is built up, which assists the retention of a high proportion of water. In a deflocculated system (a less common one for plastic clays) the particles of clay are less aggregated, there is less internal structure and so much less water can be entrapped.

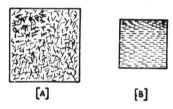

Fig. 56. Packing of particles in (A) Ca-clay, (B) Na-clay.

It follows, therefore, that a Na-clay requires much less water to make it flow readily than does a Ca-clay. The situation inside plastic masses of Ca-clay and Na-clay can be visualised from Fig. 56. In (A), the Ca-clay, the particles are strongly agglomerated and the packing of particles is random and very loose, with a high void space for entrapment of water. Such a mass has a high yield value owing to the internal structure and requires a high water content to enable it to flow. By contrast, the Na-clay consists largely of discrete particles, since agglomeration is inhibited by the high energy of repulsion. The packing of the particles is much denser than with the Ca-clay and so much less water can be entrapped. Accordingly, very little water is required for flow and the water tolerance is low. The contrast in the behaviour of the two systems is shown in Fig. 57, where the pressure–flow curves of the two clays are compared at the same water content. Whilst the sodium clay attains maximum flow rate at under 10 lb in^{-2}, the calcium clay requires nearly 70 lb in^{-2} for the same flow rate. Other significant features are the high yield value of the Ca-clay and the corresponding greater

The Plasticity of Clays

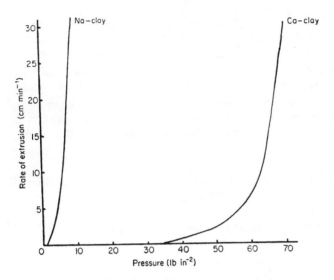

Fig. 57. Effect of exchangeable ions on the flow curve.

'bottom curvature', both associated with the strong internal structure which is only gradually broken down by shear.

Comparison of the plasticities of the various clay minerals

The expanding-lattice minerals show a marked difference in behaviour from the non-expanding minerals. In montmorillonites, for example, water can penetrate the lattice and in consequence more water is required to render montmorillonites workable than kaolinites. When well dispersed, the montmorillonites tend to break down progressively, and in theory could approach unit cell size, though it is doubtful whether this is ever achieved in practice. Such extensive breakdown increases the effective surface area and accordingly the montmorillonites exhibit typical clay properties such as plasticity and thixotropy to a greater degree than kaolinites, as indicated by the Atterberg values (Table 24). The flow curves of montmorillonites are characterised by high yield values and high flow rates, the curves being very steep (Fig. 51).

The plasticity of the kaolin minerals is of a lower order than that of montmorillonites. Nacrite and dickite appear to possess very little plasticity, possibly because their particles are large and 'chunky'. Well-crystallised kaolinite, as occurs in china clay, has a moderate degree of plasticity, but the disordered kaolinites occurring in British sedimentary clays are much more plastic, being intermediate between china clay and montmorillonite; this probably reflects the higher proportion of fine particles present in the disordered kaolinites. It should be remembered, however, that naturally occurring clays almost invariably contain non-plastic accessory minerals that reduce the overall plasticity of the clay. Little seems to be known of the plasticity of halloysite, but reported Atterberg values suggest that it may be similar to that of kaolinite.

APPLICATION OF THE RHEOLOGICAL AND OTHER PROPERTIES OF CLAY

Filter-pressing and other dewatering processes

Attention has previously been drawn to the packing of the particles in flocculated and deflocculated clay masses. In the pottery industry, raw materials are usually wet-mixed for efficiency and the beneficial effect on plasticity; a great deal of excess water is inevitably used and this must be removed to convert the clay into a plastic state for shaping articles. The necessary dewatering is accomplished by filter-pressing, a process in which the slurry is filtered under pressure through fine filter-cloths. Press-cakes formed by this process have considerable thickness and it is clearly the permeability of these press-cakes to water that determines, along with the pressure, the rate of filter-pressing. In order to obtain high permeability, the clay is not deflocculated before mixing, hence the particles pack loosely and enable water to penetrate readily. Research has shown that if the naturally occurring calcium ions are exchanged for sodium, filter-pressing becomes very slow or even ceases altogether.

Filter-pressing has been used in disposing of clay-type impurities of fine coal. In this instance, the clay may initially be in a deflocculated state because of the presence of reagents used in froth-floating the coal; to facilitate filter-pressing, it is then necessary to add various flocculating agents.

Slip-casting

The method of controlling the quality of a casting-slip by adjustment of its rheological properties has already been discussed. However, although the presence of a yield value and sufficient fluidity are important factors in themselves, the underlying reason for the dependence of the quality of casts on the rheology of the slip is the precise manner in which the clay particles are arranged in the cast as it is formed.

Immediately adjacent to the wall of the plaster mould, water is abstracted from the slip so that the latter increases in solid content to form a more-or-less rigid layer of clay adhering to the wall. Although the mould may be very porous and has a considerable suction pressure, from this moment the rate at which water is removed from the slip depends principally on the permeability of this layer of clay. Clearly, for maximum rate of water removal, a flocculated system would appear preferable, because of

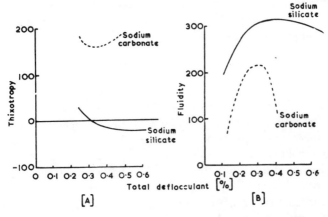

Fig. 58. The effect of deflocculants on (A) thixotropy and (B) fluidity.

its high permeability; however, this is more than offset by the much higher water content needed by a flocculated system. If the system is fully deflocculated, however, the casting may then be unacceptably slow and moreover mould release may be poor because of the small drying shrinkage of a fully deflocculated clay. Hence, a compromise is effected by adjusting the electrolyte to just below that required for optimum deflocculation. In these circumstances, some residual thixotropy is always present, and indeed desirable, because of its indication of residual structure in the slip, which in turn indicates a measure of permeability. Too high a thixotropy is equally to be avoided, of course, since this could cause the entire slip to acquire a temporary rigidity in the mould, resulting in an unduly thick, flabby cast. Figure 58 shows the effect of two different electrolytes on the thixotropy and fluidity of a casting-slip.

The total permeability of the cast at any instant, and for a constant packing density, is inversely proportional to its thickness; hence, the rate of casting is also inversely proportional to the thickness. The same concept is expressed in an equation used by Herrmann and Cutler:

$$\frac{dL}{dt} = KD \cdot \frac{\Delta C}{L}$$

where L = thickness of slip-cast wall at a time t from the commencement of casting, D = diffusion coefficient for the water passing through the cast layer, K is a conversion factor dependent on the density of the slip, and ΔC is the difference in water concentration across the wall of slip-cast material. If L and t are the only variables, direct integration of this expression results in the relationship:

$$L = [2KD \cdot \Delta C \cdot t]^{1/2}$$

Thus, the amount of cast formed is proportional to the square root of the time. The most significant constant in this equation is the diffusion coefficient D, which clearly depends directly on the permeability of the cast layer. Experimental results on the whole agree well with this equation, suggesting that over the range of thicknesses and concentrations normally used, it is correct to assume that the permeability is constant throughout the thickness. This theoretical treatment ignores the possibility of interaction of calcium ions from the mould with the clay slip, which could lead to cation exchange and local flocculation, thus assisting the casting process; however, the fact that the cast thickness is in practice closely proportional to the square root of time indicates that such effects, if they exist, must be negligibly small.

Extrusion

In the extrusion machines most frequently used in the ceramic industry, clay is forced along a barrel by means of a screw which rotates at a constant but controlled speed. The barrel tapers at the exit end and emerges via an appropriately shaped die. The flow inside the extruder is much more complicated than the simple case discussed earlier, where a pressure is applied directly; this is because the pressure built up in the extruder as the screw forces the clay in a forward direction parallel to the axis causes some to flow in the reverse direction through leakage between the tips of the screw flights and the barrel. Moreover, the flow pattern is distorted at the exit end by the taper of the barrel and by the die. In an experimental study of extrusion, Goodson found that the flow was related to the speed of the screw shaft by the parabolic equation:

$$Q = AN - BN^2$$

where Q = rate of flow, N = shaft speed, and A and B are constants. In this equation it is suggested that the first term represents the forward flow of the clay, which is directly proportional to the shaft speed, and the second term the reverse flow, which presumably is proportional to the square of the speed. It is clear that Q will have a maximum at some value of N, beyond which the flow actually decreases as the speed N is increased further. Although the equation predicts a negative (i.e. reverse) flow at very high speeds, the experiments carried out covered a range from zero up to a value of Q at or around the maximum.

As would be expected of a body having a yield value, it is found that no extrusion occurs until a certain pressure is exceeded. Goodson's theoretical treatment of extrusion was strictly applicable to Newtonian fluids only, but more recently Capriz has published a theoretical analysis of extrusion processes which takes account of non-Newtonian behaviour.

From time to time, various claims have been made concerning the beneficial effect on extrusion of various additives. Sodium carbonate has been added, for instance, to reduce the amount of tempering water required for extrusion. Whilst such additions appeared to be of benefit for certain poorly plastic shales, in other cases there was little advantage; the situation is thus analogous to that of drying. Again, surface-active agents that lower surface tension have been added. These substances are believed to coat the surface of clay particles with organophilic groups, thus rendering it more compatible with paraffinic lubricants. The result is to enable extrusion to be performed at lower pressures. However, there is no single combination of additives that is effective for all types of clay, but this is a subject that probably warrants further detailed study.

Drilling fluids in the petroleum industry

In drilling for petroleum, a drilling fluid or mud is constantly circulated from around the drill to facilitate removal of cuttings and entrained gas. The drilling fluid is pumped from the drill hole into a pit, where cuttings are removed by sedimentation, and then fed back to the drill. The fluid used needs to have a reasonably low viscosity for easy circulation but also requires sufficient thixotropy to prevent settling of cuttings in the drill hole, thus causing the drill to 'freeze'. The thixotropy should not be so high as to prevent subsequent removal of cuttings and entrained gas when the fluid is pumped out.

Evidently, control of the rheological properties of the drilling fluid is vital, yet, as with casting-slips, a compromise in properties has to be achieved. For this purpose, certain montmorillonites (notably Wyoming bentonite) and sepiolite have proved very effective. As with casting-slips, adjustment of viscosity and thixotropy by the addition of electrolytes may prove necessary, particularly when drilling through beds of gypsum or other strata likely to cause changes in consistency through ion exchange.

Civil engineering

Since clay is one of the principal constituents of soil, its behaviour when under stress is of very great importance. This is true whether one is considering the construction of a building or of a tunnel or cutting. Before any constructional work is commenced, it is the practice to sample the underlying soil by drilling out a core, on which a number of tests are carried out. Among these are determination of Atterberg limits, compressive strength, shear strength and permeability. Another quantity that may be useful is the *liquidity index*, which is defined as:

$$\text{Liquidity index} = \frac{\text{'Natural' water content} - P_w}{L_w - P_w}$$

The 'natural' water content referred to above is that present in the soil at the time of sampling; L_w and P_w are the liquid and plastic limits defined previously, and the denominator is obviously the Atterberg plasticity index. For a liquidity of unity, the natural water content must obviously be equal to the liquid limit. The liquidity index has been related to the consolidation of the soil under pressure.

The mineralogy of the soil is also worthy of investigation. Although it is scarcely practicable to make an accurate prediction of all the physical properties of a soil from its mineralogical constitution, a knowledge of the latter may enable one to predict unusual behaviour in a soil having an uncommon mineralogical composition.

The use of clay in the paint industry

Originally used as inert fillers to reduce cost, clays are now essential ingredients in paints, to which they impart many desirable properties.

Kaolinites of controlled particle size are included in paints to assist dispersion of the pigments, thus enabling high solid concentrations to be used; they may also increase hiding power. For oil-based paints, clays coated with cationic surfactants (such as amines) are used to make them organophilic and compatible with the liquid. Bentonites are also used in a similar fashion. Clay additions may be used to impart a degree of thixotropy to a paint, giving it 'non-drip' qualities. Amounts used are from 0·2 to 1 kg/litre.

Use of clay in the paper-making industry

Various inorganic substances are incorporated into paper as fillers, to give opacity and affinity for printing-ink; and as coatings, to give a gloss, whiteness and good ink receptivity. Clays are used extensively for this purpose, some 70% of china clay production in the United Kingdom being used for paper. China clay only is used for this purpose; special 'paper grades' of the clay are produced, of which the physical properties are carefully controlled. Particle size, whiteness, opacity and good retention on the wire of the paper-making machine are the important properties for filling clays, whilst, in addition, coating clays are required to have suitable rheological properties and a good gloss.

The general requirement is for a clay that can be suspended in water at a concentration of some 70% (w/v) with the aid of a deflocculant. At this concentration, the suspension must be fluid enough to pass readily through fine screens and to allow rapid spreading and levelling on the paper surface by machine. A small degree of thixotropy is desirable, but under no circumstances should the suspension become dilatant. The rheological properties are thus quite critical and careful control is necessary during production. Moreover, viscometric measurements are carried out by the clay producer to ensure that the various grades of clay meet the requirements. The presence of montmorillonite as an impurity in some china clays may cause difficulties, since the former tends to concentrate in the fine fractions and may result in the viscosity becoming too high.

Clay soils in agriculture

Clays in general are important constituents of the soil. The most common clay mineral of soils is kaolinite, frequently the disordered type; but montmorillonite and illite also occur in significant amounts. Both the chemical and physical properties of the clay component of soil have a direct bearing on the plant life which the soil supports.

Some of the earliest work on the ion-exchange properties of clays was done by soil scientists, who first reported the uptake of bases by clay soils. Lime is a commonly used additive for soils, and apart from its beneficial effect on the texture, the resultant calcium ions replace the existing ions of the soil, releasing them in a soluble form and rendering them available to plants. The retention of potassium and of phosphate have also been studied from the point of view of plant nutrition.

A further effect of cation exchange is of course its effect on the rheological properties of soil. The addition of lime converts the clay soil to the calcium form which, as already described, has a very permeable texture, capable of retaining a considerable amount of water without becoming waterlogged, yet capable of releasing the water when external conditions become relatively dry. A sodium clay soil, by contrast, is comparatively impermeable to water, thus preventing it escaping readily by percolation; moreover, because of its high packing density, it is heavy and intractable. As an alternative to lime, certain organic reagents have been used for conditioning soil; in these instances it is probable that the effect of these substances is to produce a highly flocculated soil, similar to a calcium type but having a greater void space.

READING LIST

N. F. ASTBURY, *Trans. Brit. Ceram. Soc.*, **60** (1961) 1.
E. BUCKINGHAM, *Proc. Amer. Soc. Test. Mater.*, **21** (1921) 21.
G. CAPRIZ, *Trans. Brit. Ceram. Soc.*, **62** (1963) 339.
F. J. GOODSON, *Trans. Brit. Ceram. Soc.*, **58** (1959) 158.
R. E. GRIM, *Applied Clay Mineralogy*, McGraw-Hill, 1962.
E. R. HERRMANN, and I. B. CUTLER, *Trans. Brit. Ceram. Soc.*, **61** (1962) 207.
H. H. MACEY, *Trans. Brit. Ceram. Soc.*, **43** (1944) 16.
C. E. MARSHALL, *Colloid Chemistry of the Silica Minerals*, Academic Press, 1949.
F. MOORE, *J. Sci. Inst.*, **40** (1963) 228.
F. MOORE, *Rheology of Ceramic Systems*, Institute of Ceramics Textbook Series, Elsevier Applied Science Publishers, 1965.
F. MOORE, and A. DINSDALE, *Viscosity and its Measurement*, Institute of Physics, 1962.
V. J. OWEN, and W. E. WORRALL, *Trans. Brit. Ceram. Soc.*, **59** (1960) 285.
M. REINER, *J. Rheology*, **1** (1930) 251.

Chapter 8

The Effect of Heat on Clays

THE DRYING OF CLAYS

Critical moisture content

The drying of clays and clay-based ceramics is of considerable technological importance. When a plastic clay is dried, shrinkage occurs; if drying is too rapid and uneven, the shrinkage may give rise to cracks in the product. When completely dry, clays have a considerable strength which is further enhanced by firing.

Consider a mass of plastic clay which is being allowed to dry (Fig. 59A). As drying proceeds, water evaporates from the outer surface and the

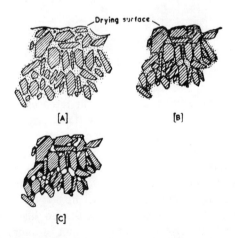

Fig. 59. Behaviour of clay particles on drying.

particles of clay are drawn progressively closer together by surface tension forces. Eventually, the clay particles will come into mutual contact (B), forming a loosely packed assembly. When this stage has been reached, further contraction is not possible and no further shrinkage therefore occurs. The residual water is contained in the voids between the particles and the water content at this stage is known as the *critical moisture content* (c.m.c.). Further drying now results in loss of water from the pores of the body, the water being drawn to the surface by capillary attraction. Thus,

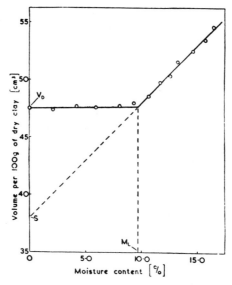

Fig. 60. Drying curve for a clay body.

the original packing is maintained and air replaces water in the pores, resulting finally in a dry, porous body (C). If measurements of volume and moisture content are made periodically during the drying and plotted as shown in Fig. 60, two straight lines are obtained, connected in most instances by a small curved portion. The upper part of the graph represents the stage where water is being lost and the reduction in volume corresponds exactly to the volume of water, i.e. the body shrinks 1 cm^3 for each 1 g loss in weight. This initial stage has been called the *constant rate period*. At the c.m.c., which corresponds to the point of intersection of the two lines, shrinkage ceases but weight loss still occurs as the graph shows. This second stage, where water is being lost by diffusion through the pores, has been termed the *falling rate period*. In practice, it has been found that the graph is not absolutely straight but bows very slightly so as to be convex to the moisture axis, implying that there is a small shrinkage beyond the critical point, but this is followed by a slight expansion that returns the volume to its original value at the c.m.c. This minor effect has been attributed to a slight relaxation effect that occurs in the last stages of drying.

Although a well-defined drying curve is readily obtained with small laboratory test-pieces, slightly different results may be obtained with larger articles such as building-bricks because the moisture content is then no longer uniform. The moisture content at which shrinkage first ceases is then less sharply defined and is referred to as the *leather-hard moisture content*.

Factors affecting the rate of drying

During the constant rate period of drying, there is an unbroken film of water on the surface of the clay body and therefore the rate of drying is governed by evaporation from a free surface; it thus depends on the amount of surface, on the temperature and on the humidity of the surroundings, whether the latter applies to open-air drying or to the interior of a commercial drier. Beyond the c.m.c., however, there is no longer a continuous film covering the surface, hence the maximum rate of drying can no longer be maintained; during this second stage, then, the rate of drying depends also on the rate at which water can diffuse through the pores to reach the surface.

Ford (1964) has found that the ease of drying of naturally occurring clays is closely linked to their mineralogical compositions; those clays consisting chiefly of kaolinite can be dried safely at much higher rates than those consisting of illite. Recently, Al-Khalissi and Worrall (1985) have investigated the pore structures of a number of dry, unfired clay specimens by mercury porosimetry. The results show that the kaolinitic clays have the largest average pore diameters, illitic clays the smallest. These findings clearly accord well with those of Ford, since the larger pores of the kaolinitic clays facilitate the rapid diffusion of water throughout the specimens, thus enabling them to be dried rapidly without risk of cracking.

The precise reason for this difference between kaolinites and illites is largely a matter of conjecture. Work by the author and his associates indicates that there is no relationship between pore size and particle size distribution. There remains the possibility that the modes of packing of the particles differ in some way. As mentioned previously (Chapter 5), the edges of kaolinite particles can acquire a positive potential under some circumstances, resulting in electrostatic attraction between edges and faces; this type of association leads to larger pore diameters than would occur with any other type of particle–particle packing. Illite on the other hand may be less prone to this kind of particle packing because of the different character of the crystal edges but this possibility needs to be confirmed by experiment.

Various additions have been made to clays in order to accelerate drying without at the same time causing disruption of the body through the creation of strains in its interior. Chemical additives have in general been flocculating agents such as calcium chloride and hydrochloric acid. Where these additives have been effective they have probably functioned by increasing the degree of flocculaton of the clay, thereby making the texture more 'open', and thus enabling the moisture to escape more readily from the interior. They are not effective in all cases, possibly because of secondary reactions with accessory minerals or because the clay is already highly flocculated. The presence of non-clay material such as quartz also has the effect of opening up the texture and thus facilitating drying, but although much quartz may be naturally present in a clay, addition of this material may not be practicable because of the inevitable reduction in plasticity that it causes.

Factors affecting the critical moisture content and total shrinkage

As mentioned previously, flocculated systems form loosely packed masses of low density, whilst deflocculated systems, in contrast, pack much more densely; accordingly, the critical moisture content of a flocculated system will be higher than that of a deflocculated system, because the former has a much greater void space when it reaches the critical stage and thus contains more entrapped water (cf. A and B in Fig. 56). Thus, flocculating agents increase the c.m.c. whilst deflocculating agents decrease it. The relationship between the initial moisture content M_0 and the critical moisture content M_c during the constant rate period can be readily deduced as follows. For simplicity, suppose that the true density of the clay is $2 \cdot 5$ g cm^{-3} and that of water $1 \cdot 0$ g cm^{-3}, and that a given specimen contains 100 g of clay. The loss of water in drying from M_0 to M_c will be $(M_0 - M_c)$ g and so the volume is $(M_0 - M_c)$ cm^3. During the constant rate period, this volume represents the shrinkage of the specimen. Now the initial volume of the specimen = (volume of water) + (volume of clay) = $M_0 + 100/2 \cdot 5 = M_0 + 40 \, \text{cm}^3$. Hence the total shrinkage expressed as a percentage of the original volume is:

$$\frac{(M_0 - M_c) \times 100}{(M_0 + 40)}$$

Under ideal conditions, therefore, the total shrinkage depends on the difference between the critical moisture content M_c and the initial moisture content M_0. The latter is adjusted in practice to suit the particular forming process so as to produce the correct consistency. Evidently, deflocculants could be used to reduce this initial moisture content whilst maintaining correct consistency, but the shrinkage is not necessarily reduced because the c.m.c. is also reduced. On the other hand, if flocculating agents are added to increase the c.m.c., this is compensated by an increase in the initial moisture required and again the shrinkage is but little affected. In order to alter substantially the total drying shrinkage it would be necessary to *increase* the c.m.c. and to *decrease* the initial moisture content; this might possibly be achieved by the addition of a weak deflocculant that was unstable and decomposed during drying or, better still, one that became a flocculant as drying progressed.

Linear shrinkage

The foregoing discussion has been concerned with *volumetric* shrinkage but the *linear shrinkage* must also be considered in industrial practice. For volumetric shrinkages of 10% or less, the linear shrinkage is very approximately one-third of the former for an isotropic specimen, i.e. a body that shrinks equally in all directions. However, this is seldom true of clay-based articles because the plate-like clay particles have a strong tendency to assume a face-to-face orientation over certain regions; this is particularly likely to occur at the walls of the mould for slip-cast articles, or parallel to the sides of the die for extruded bodies, for example. Within such regions of strong orientation, the linear shrinkage will clearly differ according to whether it is measured parallel to the plates or perpendicular to them.

The dry strength of clay articles

It is well known that clay bodies when thoroughly dried have a considerable strength. This strength is maintained up to the temperature of decomposition of the clay and hence is not dependent on the presence of adsorbed water films. Nevertheless, it is significant that clay is not rendered plastic and develops no unfired strength when mixed with non-polar liquids. It is probable, therefore, that strongly adsorbed, stable films of water are essential for the development of unfired strength and plasticity. It also seems likely that hydroxyl groups providing hydrogen bonds must also be present, since non-hydroxylic liquids, even if polar, are not plasticisers. It is likely that as the clay dries, the adsorbed liquid films contract and the surface tension forces pull the clay particles together, ensuring maximum area of contact. When the last traces of water have been expelled by drying, the electrostatic forces of repulsion cease to exist and the only forces then available are the short-range van der Waals type which are presumably responsible for the dry strength. The action of water or other plasticiser in bringing the particles of clay into intimate contact is essential in promoting strength; a perfectly dry body could not develop such high strength merely by pressure.

This does not mean, however, that the dry strength of a given clay is constant. Normally, clays used in plastic forming are flocculated, but if a deflocculant is added, the particles no longer exist as loose aggregates but remain discrete until the system is almost dry. Then, however, they pack very densely, with greater area of contact and hence increased strength. This does not imply that it would be beneficial to deflocculate plastic clays industrially, because such systems are very sensitive to small changes in moisture content and are thus difficult to control, and to dry.

It is thus somewhat paradoxical that deflocculated systems, which are associated with low viscosity and low 'wet strength' (yield value), eventually dry to give bodies of relatively high strength compared with flocculated systems.

THERMAL DECOMPOSITION OF CLAYS

Pure clay minerals

The effect of heat on pure kaolinite has attracted a great deal of attention. From the earliest days, it has been known that, after the removal of adsorbed water at just over 100°C, kaolinite decomposed above about 450°C, losing its hydroxyl groups as water. Work done by White and others has shown that this dehydration follows approximately a first-order reaction.

The precise nature of the first product of dehydration, however, has been a matter of dispute. Formerly believed to be a mixture of amorphous alumina and silica, the residue has now been shown to possess some definite structural features, some of the original kaolinite structure persisting. Moreover, if the temperature has not been too high, the residue can be rehydrated to kaolinite. Accordingly, it has become current practice to refer to this product as *metakaolin*:

$$Al_2Si_2O_5(OH)_4 \xrightarrow{450°C} Al_2Si_2O_7 + 2H_2O$$
$$\text{kaolinite} \qquad\qquad \text{metakaolin}$$

With disordered kaolinites, the dehydration product is somewhat less well defined.

The dehydration of kaolinite may be followed by differential thermal analysis, an endothermic peak occurring, with the standard rate of heating, at 600°C for well-crystallised kaolinite and at about 580°C for disordered kaolinite.

At still higher temperatures the metakaolin undergoes further reactions to form crystalline compounds, the end-products being free silica (cristobalite) and mullite, $3Al_2O_3.2SiO_2$. Thus, it is clear that there occurs a separation into an alumina-rich compound and free silica, but the precise manner in which this takes place is still a matter of controversy.

Earlier work, based chiefly on X-ray diffraction, indicated that at 1000°C a spinel-type compound was formed, thought to be γ-Al_2O_3. At this temperature, thermal analysis shows an exothermic peak, also thought to be associated with the recrystallisation of γ-Al_2O_3. Other authorities, however, attributed this exothermic peak to the formation of mullite; but this now seems unlikely, since recent X-ray work strongly suggests that mullite is not formed until much higher temperatures are attained (1150–1300°C). It has been reported, from experiments with silica–alumina gels, that sillimanite, $Al_2O_3.SiO_2$, a compound structurally similar to mullite, can be formed at temperatures as low as 1000°C, but this might be because the silica–alumina gels are much more reactive than metakaolin and moreover may even retain a proportion of their combined water at a high temperature.

More recent accurate measurements of the unit cell dimensions of the spinel compound have shown that it is probably not γ-Al_2O_3 but a silicon-aluminium spinel, of formula $2Al_2O_3.3SiO_2$. This latter compound is then said to lose silica by a progressive diffusion of Si ions from the lattice, forming a mullite-like compound, $Al_2O_3.SiO_2$, as an intermediate stage, and by further loss of silica, true mullite, $3Al_2O_3.2SiO_2$, is formed. The entire process may therefore be represented by the following chemical equations:

$$2(Al_2O_3.2SiO_2) \xrightarrow{925°C} 2Al_2O_3.3SiO_2 + SiO_2$$
metakaolin $\qquad\qquad$ silicon spinel

$$2Al_2O_3.3SiO_2 \xrightarrow{1100°C} 2(Al_2O_3.SiO_2) + SiO_2$$
silicon spinel $\qquad\qquad$ pseudo-mullite

$$3(Al_2O_3.SiO_2) \xrightarrow{1400°C} 3Al_2O_3.2SiO_2 + SiO_2$$
pseudo-mullite $\qquad\qquad$ mullite $\quad$ cristobalite

The liberated silica, it will be noted, forms cristobalite, but this transformation is commonly believed not to occur below 1350°C. However, this mechanism, first proposed by Brindley and Nakahira, is the currently accepted one. There is some evidence that disordered kaolinite behaves in a slightly different manner to that described, because of its lower crystallinity and the presence of ions other than aluminium and silicon in the structure.

The effect of impurities

Naturally occurring clays in general contain minerals other than kaolinite, the most common being quartz and mica, although a large number of minor constituents, notably pyrites (FeS_2), oxides of iron, calcite ($CaCO_3$), dolomite ($CaMgCO_3$), gypsum ($CaSO_4.2H_2O$) and anatase (TiO_2) have also been reported. One of the most important effects of impurities is that due to fluxing ions, namely Na, K, Ca and Mg, derived from the micas and other compounds. These ions when present lower the melting point of the system and, in particular, react with silica at as low as 1000°C to form a viscous liquid, which on cooling does not crystallise but solidifies to form a glass, in which the fluxing ions occupy holes in the Si—O— network. This complex glass is responsible for the vitrification and fired strength of such clays. Quartz in naturally occurring clays is coarse-grained and therefore reacts but little, often appearing unchanged after heating; any that is fine enough to react is probably incorporated in the glassy phase. From the foregoing discussion, it will be appreciated that the principal products of fired clays are mullite, a complex glass, cristobalite (if undissolved by the glassy liquid) and possibly some unchanged quartz.

The other kaolin minerals, halloysite, dickite and nacrite, have all been reported to develop spinel-type phases when heated above 1000°C, but the mechanism of these changes has not been so thoroughly investigated as has that of kaolinite.

Thermal decomposition of montmorillonites

Because of the many different substitutions in montmorillonites, there is a wide variation in their chemical compositions and therefore in the final products of heating. However, as with kaolinite, adsorbed water is lost at about 105°C from the external surfaces, but a much higher temperature, from 120 to 300°C, is required to remove the inter-layer water, depending somewhat on the nature of the exchangeable cations. At about 650°C the hydroxyl groups are removed as water, the exact temperature varying with the type of montmorillonite. The initial dehydration products in this case are probably amorphous alumina and silica. On heating to still higher temperatures, a spinel phase appears, and the final products of heating to 1400°C are again mullite and cristobalite. The presence of fluxing ions will of course modify these reactions. If magnesium is present, cordierite and possibly periclase may be formed, as well as mullite and cristobalite. The montmorillonite group of minerals may be considered to be derived from pyrophyllite (or talc) by substitution, and if pyrophyllite is taken as a 'model' substance, the dehydration reaction may be written:

$$3[Al_2Si_4O_{10}(OH)_2] \xrightarrow{650°C} 3Al_2O_3 . 2SiO_2 + 10SiO_2 + 3H_2O$$

Dehydration of the micas

Data on the dehydration of muscovite are somewhat conflicting. Some authorities have reported a gradual loss of water at temperatures up to about 800°C, whilst others maintain that there is a period of relatively rapid dehydration between 450 and 850°C. Similar results have been reported for phlogopite and biotite. On heating muscovite to 1000°C the products are said to be γ-alumina or spinel and at 1400°C the products are reported to be α-alumina and glass.

Illites, particularly those present in British fireclays, have dehydration characteristics remarkably similar to those of kaolinite. Thus, there is a small loss of water below 100°C and a major evolution of water between 350 and 600°C. However, Grim and Bradley have reported that the structure of illite is not completely destroyed until a temperature of 850°C is attained. The high-temperature products are said to be similar to those of muscovite.

REFERENCES

AL-KHALISSI, F. Q., and WORRALL, W. E., *Br. Ceram. Trans. J.*, **84** (1985) 178.
FORD, R. W., *Drying*, Institute of Ceramics Textbook Series, Elsevier Applied Science Publishers, 1964.

READING LIST

W. F. FORD, *The Effect of Heat on Ceramics*, Institute of Ceramics Textbook Series, Elsevier Applied Science Publishers, 1967.
R. E. GRIM, *Clay Mineralogy*, McGraw-Hill, 1968.

Chapter 9

Methods Used for the Identification and Characterisation of Clays

CHEMICAL ANALYSIS

The techniques of chemical analysis are fully described in specialised books, a list of which is included at the end of this chapter. The purpose of the present discussion is to indicate the uses and limitations of chemical analysis as applied to clays.

One of the disadvantages of chemical analysis was formerly the great length of time required for its completion. The 'classical' method, as it is called, gives considerable accuracy but skill is required for reliable results. Recent developments have favoured instrumental techniques which, once properly calibrated, do not require as much skill as the classical method, and are moreover much less time-consuming. In the *flame photometer*, for example, a suitably diluted solution of the sample is sprayed into a flame thus producing the characteristic spectrum of the more strongly emitting elements, notably the alkali metals. An appropriate wavelength is selected, either with filters or a monochromator, and the intensity of the radiation is measured with a photocell or photomultiplier. With town gas and air, only the alkali metals and sometimes calcium can be determined; the use of hotter flames (e.g. air–acetylene) enables other elements such as magnesium to be detected, but at these higher temperatures interference becomes troublesome and largely outweighs the advantages of increased sensitivity.

It is probably best to use emission techniques for the alkali metals only, and to determine the heavier elements where appropriate by atomic absorption. In this technique, a solution of the sample is sprayed into a flame as for emission; monochromatic light of appropriate wavelength is passed through the flame and the reduction in intensity due to absorption is measured by means of a sensitive photomultiplier. The concentration of the element in question is approximately proportional to the absorption, but in practice the instrument is always calibrated against standard solutions. Since each element absorbs radiation at certain frequencies only, the method is very selective and relatively free from spectral interference. Moreover, the sensitivity is high because absorption is due to atoms in the *ground state* (only a minute fraction of the atoms in the flame are excited sufficiently to emit, the remainder being in the ground state). Whilst excellent for minor constituents, atomic absorption is scarcely precise enough for major constituents (aluminium and silicon); moreover, the preparation of the aqueous solutions is time-consuming, although less so than for classical analysis.

Colorimetric analysis is still used for elements such as iron that give strong colours when reacted with a suitable reagent. At the present time the intensity of the coloured solution is invariably measured with some type of spectrophotometer, based on a monochromator for wavelength selection and a photosensitive device for detection. The method is very suitable for the determination of a few per cent of iron or titanium that is commonly present in clays; it is not, however, as wide-ranging as atomic absorption.

Another technique that is being increasingly employed is that of *X-ray fluorescence*. The sample is either pressed in powder form into a compact or, better still, fused with a suitable flux (e.g. borax) to form a glass disc. A thin pencil of X-rays from a tungsten or chromium tube is allowed to fall on the specimen; this causes elements of lower atomic number to be excited and produce secondary or fluorescent X-rays which are characteristic of the elements in the sample. A suitably strong spectral line from a given element is selected by means of an analysing crystal and the intensity of the corresponding radiation is measured with a scintillation or flow counter, depending on the wavelength. Calibration against a series of standards is of course essential but is complicated by interferences of various kinds. The most serious of these is that the characteristic radiation of any given element is absorbed to some extent by other elements, so that extensive corrections need to be made. Nevertheless, the technique is promising and, unlike most instrumental methods, can give fair precision on major constituents.

It is likely therefore that instrumental methods will replace classical methods for routine work, but the latter will still need to be used for occasional checking and calibration.

Purification of clays

The interpretation of chemical analyses is simpler when a pure sample of the relevant clay mineral is available. Various methods are known for the separation of various minerals, but few of them are effective for clays. One most frequently used is that of sedimentation, which aims at separation of minerals on a size basis; since no two minerals are likely to have exactly the same size distribution, this method is often effective for the separation of clay from sand, silt and quartz. Clay minerals, for instance, usually consist of particles less than 1 μm in diameter, whilst quartz has particles of diameter from 1 μm upwards. Therefore, if a size 'cut' is made at 1 μm or, better still, at 0·25 μm, the material less than the specified size will be free from quartz. Other coarse-grained minerals that can be removed from clay by this means are calcite, pyrites and hydrated iron oxide. In practice, several kilograms of the raw material are dispersed in an excess of water (preferably using a volatile dispersing agent) and the suspension allowed to settle for a suitable period of time. According to Stokes' Law:

$$r = \left[\frac{9\eta h}{2(d_1 - d_2)gt} \right]^{1/2}$$

where r is the radius of a given particle (assumed spherical), h the height through which it falls in time t, g the acceleration due to gravity, d_1 and d_2 the densities of the solid and liquid respectively, and η the viscosity of water. From this equation, the time required to sediment all particles at depth h (usually measured from the surface of the suspension to the bottom of the container) can be calculated. It is on this principle that the method of Andreasen and Lundberg (1930) for the determination of particle size distribution is based. For particles of radius down to 1 μm gravitational sedimentation is satisfactory, but below 1 μm it is necessary to employ a centrifuge, in which g is replaced by a centrifugal force that may be a thousand times g or greater.

For completely unknown mixtures of minerals, and for all precise work, several size 'cuts' should be made, thus obtaining a series of fractions, each falling within a narrow size range. Thus, any variation of chemical composition and other properties from one size range to another can then be detected. If a number of successive fractions have constant properties, this is evidence (though not absolute proof) that a pure mineral species has been isolated.

In addition to size fractionation, miscellaneous methods may be used in conjunction with the latter. In the method of heavy liquid separation, for example, a series of mixtures of bromoform and benzene are made up, to give a range of densities. These mixed liquids can be adjusted so that certain minerals just sink or float when placed in them. Unfortunately, heavy liquid separations are generally ineffective for particles of submicron size, because of the difficulties of dispersion.

In froth flotation, use is made of the selective adsorption of an organic molecule on to a specific mineral, rendering the latter hydrophobic so that it is repelled by the aqueous phase and collects at the air–liquid interface, usually assisted by the production of a froth. This method, too, is inefficient with submicron particles. In electrostatic and magnetic separation, use is made of differences in dielectric constant and magnetic susceptibility, but these methods again are only effective for particles greater than several microns in radius.

It will be appreciated from the above discussion that it is generally rather difficult to establish the absolute purity of a specified mineral. The final product should in all cases be examined by as many techniques as possible, the most important being X-ray diffraction, differential thermal analysis and infra-red absorption.

'Rational' and 'proximate' analyses

A number of methods have been proposed for the determination of the mineralogical composition of a clay by selective chemical attack. Mostly, these have been based on the destruction of the 'clay mineral' by heating with strong sulphuric acid, leaving the 'mica' and quartz unattacked. The separation is generally unsatisfactory, because the clay mineral is not completely decomposed and moreover mica is frequently attacked also. These methods come under the general heading of *rational analyses*. Alternatively, the mineralogical composition may in favourable circumstances be calculated from the ultimate chemical analysis, giving the so-called *proximate analysis*. The term *rational analysis* has unfortunately been used synonymously with proximate analysis and this terminology has persisted. The calculation of mineralogical composition from chemical data is therefore commonly referred to as a rational analysis.

The calculation requires a number of assumptions, the main one being that the minerals present have already been determined qualitatively by X-ray, thermal analysis, etc. A further assumption (and probably the one least justified) is that each mineral has the 'ideal' unsubstituted formula. Depending on the complexity of the raw clay, the calculation can be very involved, requiring the solution of a number of simultaneous equations.

Table 25
The Chief Minerals present in Ball Clays

Mineral	Formula	Molecular weight
Kaolinite	$Al_2O_3.2SiO_2.2H_2O$	258·17
Soda mica	$Na_2O.3Al_2O_3.6SiO_2.2H_2O$	764·43
Potash mica	$K_2O.3Al_2O_3.6SiO_2.2H_2O$	796·65
Quartz	SiO_2	60·09

Table 26
Typical Analysis of a Ball Clay

Oxide	Weight (%)
SiO_2	51·13
Al_2O_3	29·30
Fe_2O_3	2·78
TiO_2	2·08
CaO	0·15
MgO	1·68
Na_2O	0·12
K_2O	2·80
Ignition loss	9·54
Total	99·58

However, for simplicity one may consider a ball clay or fireclay, where the main constituents are kaolinite, micas and quartz. The formulae assumed for these minerals, with their respective 'molecular weights', are shown in Table 25.

For the purpose of the calculation, the analysis of a typical ball clay is shown in Table 26.

The method of calculation is simplified if one commences with the alkalis. Since the 'molecular weight' of Na_2O is 61·98, it will be seen from Table 25 that 61·98 parts by weight of Na_2O correspond to 764·43 parts by weight of soda mica, so that 1 part of Na_2O is equivalent to 764·43/61·98 = 12·33 parts of soda mica. From Table 26, the percentage of Na_2O in this clay is 0·12 and the percentage of soda mica is therefore 0·12 × 12·33 = 1·48%. In exactly the same way, it is found that 1 part by weight of K_2O is equivalent to 8·46 parts of potash mica, and hence the percentage of potash mica in the clay is $2·80 \times 8·46 = 23·72\%$.

In the next step of the calculation it must be remembered that the alumina (total 29·30 from Table 26) is combined in three minerals— kaolinite, soda mica and potash mica. Knowing the amounts of the last two minerals, we can calculate the amount of alumina they contain. Subtracting this from the total alumina gives the alumina present as kaolinite, from which the percentage of the latter can be calculated. The calculated percentage of alumina in soda mica is 40·02% and so the percentage alumina combined as soda mica is $1·48 \times 0·4002 = 0·59\%$ (to two decimal places). Similarly, the percentage of alumina in potash mica is 38·40% and so the amount of alumina combined as potash mica is $23·72 \times 0·3840 = 9·11\%$. Hence, the total alumina in the form of micas is $0·59 + 9·11 = 9·70\%$. Therefore, the remainder of the alumina (that combined in the form of kaolinite) $= 29·30 - 9·70 = 19·60\%$.

From the 'molecular weight' of alumina (Al_2O_3), 101·96 parts of Al_2O_3 correspond to 258·17 parts of kaolinite, hence 19·60 parts of Al_2O_3 correspond to $19·60 \times 258·17/101·96 = 49·63\%$ of kaolinite.

Finally, the quartz or 'free silica' has to be calculated. The total SiO_2 quoted in Table 26 consists not only of quartz but of silica that is combined in the micas and in the kaolinite. Therefore, the amount of combined silica is first calculated (in the same way as for alumina) and the combined silica is then subtracted from the total silica, giving the quartz. The combined silica is found to be 34·52, so that free silica $= 51·13 - 34·52 = 16·61\%$, the percentage of quartz.

A table of useful conversion factors for rational analysis is given in Table 27.

Table 27
Conversion Factors for Rational Analysis

Constituent sought	Found as	Conversion factor
Soda mica	Na_2O	12·33
Potash mica	K_2O	8·46
Alumina	Soda mica	0·400
Alumina	Potash mica	0·384
Kaolinite	Al_2O_3	2·532
Silica	Soda mica	0·471
Silica	Potash mica	0·452
Silica	Kaolinite	0·465

The results of the above calculation of rational analysis are summarised in Table 28. It will be noted that the total of major constituents falls short of 100% by 8·59%, this latter being assumed to be miscellaneous oxides and organic matter that are not included in the calculation.

Although the results in Table 28 are quoted to two decimal places, the method scarcely warrants such accuracy, for errors arise, not in the calculation, but in the assumptions that are made. For example, the micas

Table 28
Calculated Rational Analysis

Mineral	Weight (%)
Soda mica	1·48
Potash mica	23·72
Kaolinite	49·60
Quartz	16·61
Miscellaneous oxides, organic matter, etc.	8·59

are likely to be of illite type, rather than ideal muscovite, and the kaolinite is probably of the disordered variety, all of which gives rise to errors in the conversion factors. Nevertheless, the rational analyses calculated in this way do provide a rough guide to the physical properties of the clays.

Calculation of ionic formulae

Natural minerals frequently depart from the 'ideal' formulae quoted because of isomorphous substitution, as in disordered kaolinites. In the montmorillonites, the nature and degree of substitution enables the species to be identified. It is therefore useful to have a standard method of calculating the ionic formula of a mineral from a chemical analysis, assuming the specimen to be pure. A typical analysis of a montmorillonite is given in Table 29.

As the basis of calculation, consider the 'ideal' formula:

$$Al_2Si_4O_{10}(OH)_2$$

from which all montmorillonites may be derived by substitution. In this formula, as in all others, the number of anionic or negative charges is 22, which is exactly balanced by an equal number of cationic or positive charges. The formula of the montmorillonite, the analysis of which is given in Table 29, must be derived from the 'ideal' formula by substitution of other cations for Al or Si, so that the oxygen and hydroxyl ions remain unaltered. Therefore, the number of anionic charges due to oxygen and hydroxyl is again 22, so that the number of cation charges must also be 22. The first step in the calculation, therefore, is to calculate the number of charges for each element. Considering first SiO_2, dividing by the 'molecular weight' (60·09) gives the number of 'moles' of SiO_2 per 100 g of material, and multiplying by 4 gives the total number of charges per 100 g (since Si^{4+} carries four positive charges). The result of this first step is given

Table 29
Chemical Analysis of a Montmorillonite

Oxide	Weight (%)
SiO_2	65·50
Al_2O_3	23·20
MgO	3·62
Na_2O	2·78

in column (3) of Table 30. The same operation is carried out for Al_2O_3, remembering that in this case the number of charges per unit formula is 6, and so on for the remaining elements. The number of charges for each element is added to give the result at the foot of the column (5·995). However this is merely the total number of charges per 100 g; to arrive at the number per unit formula, each figure of column (3) is now multiplied by the factor 22·00/5·995, giving the next column. Finally, to convert numbers of charges per unit formula to numbers of atoms, the figures of column (4) are divided in each case by the appropriate valency, giving column (5).

In order to present these results as a formula in the correct way, the anions are first written down, as $O_{10}(OH)_2$. The number of silicon atoms is exactly four, which is the maximum permissible in the tetrahedral layer, hence this may be written alongside the anions. The octahedral layer, containing Al, is clearly deficient in the latter, and since Mg is the only cation of appropriate size to substitute for Al in this layer, the octahedral layer is written:

$$(Al_{1·67}Mg_{0·33})$$

Table 30
Calculation of Ionic Formula

(1) Oxide	(2) Weight* (%)	(3) No. of charges per 100 g	(4) Charges per unit formula	(5) No. of atoms
SiO_2	65·50	4·360	16·00	4·00
Al_2O_3	23·20	1·365	5·01	1·67
MgO	3·62	0·180	0·66	0·33
Na_2O	2·78	0·090	0·33	0·33
	95·10	5·995	22·00	

*The weight total falls short of 100% because the combined water, which is not required for the calculation, is not included.

The remaining element, sodium, is clearly too large to be a constituent of the lattice and must therefore occupy an inter-layer position as an exchangeable cation, so that the complete formula may now be written as:

$$Na_{0.33} \leftarrow (Al_{1.67}Mg_{0.33})Si_4O_{10}(OH)_2$$

Ionic formulae for disordered kaolinites may be calculated on the same principle, the chief difference being that the total number of cationic charges is then 14 instead of 22.

CHEMICAL METHODS OF SEPARATION

Chemical methods for the separation of minerals in clays may be used either qualitatively or quantitatively and depend in general on selective decomposition, i.e. the decomposition of one or more minerals, leaving others intact. Since no chemical reagent is perfectly selective in its action, there is always some degree of uncertainty in chemical determinations of this kind. However, although the number of satisfactory chemical methods is limited, they are not necessarily less accurate than the corresponding instrumental methods. The basic principles of some of the better-known chemical methods are described below; for full experimental details, the reader is referred to the references at the end of this chapter.

Trostell and Wynne's method for the determination of quartz

A weighed sample (normally 0·5 g) of the clay is fused with some 15 g of potassium pyrosulphate ($K_2S_2O_7$) in a platinum crucible at a maximum temperature of 1000°C. By this means, the clay minerals and most other accessory minerals, apart from quartz, are decomposed. After digestion with water and sodium hydroxide solution, the mixture is filtered, washed and the residue ignited and weighed as quartz. If desired, the purity of the quartz may be checked by determination of the weight loss after treatment with hydrofluoric acid.

Despite the obvious minor errors that are possible (e.g. attack on other minerals may be incomplete) the method gives results in reasonably good agreement with those obtained by X-ray diffraction.

Hashimoto and Jackson's method for kaolinite

This procedure depends on measuring the weight loss from a sample of clay after chemical decomposition and removal of kaolinite. This is accomplished by first heating a 0·1 g sample to 500°C to decompose the kaolinite; the metakaolin so formed, being chemically very reactive, can then be readily dissolved by treatment with 0·5 M NaOH solution. After filtration and washing, the residue is weighed and the loss in weight reported as kaolinite. Whilst this procedure does appear to eliminate kaolinite completely, there is a possibility that degraded illites may also be attacked.

The dithionite method for the removal of oxides of iron

Various forms of iron oxide or hydrated iron oxide may occur in clays, sometimes as 'free' iron oxide or possibly as a thin film surrounding the particles of clay. The removal of these substances may be beneficial prior to X-ray examination or determination of cation exchange capacity. In general, the removal of iron oxides is a two-fold process; first, the iron is reduced to the (II) state, following which it is complexed in a soluble form which can then be removed by filtration or centrifugation. Mehra and Jackson (1960), for example, have recommended heating a 1 g sample of the clay to 80°C with 40 ml of 0·3 M sodium citrate and 5 ml of 0·1 M sodium bicarbonate, to which has been added 1 g of sodium dithionite, $Na_2S_2O_4$. The iron is reduced by the dithionite to Fe(II) and complexed by the citrate, whilst the sodium bicarbonate serves to maintain the pH at around 7·5. After treatment, the residual clay is freed from soluble iron by repeated centrifuging and washing with water and ethanol.

Removal of organic matter

Since the type of organic matter present in clays varies considerably according to geological age (see Chapter 4), the same method of treatment may not be satisfactory in all instances. During a study of the organic matter in ball clays, Worrall (1956) found that the former could be divided into two broad categories: (a) bitumen, consisting of waxes and resins of relatively low molecular weight; (b) lignin–humus, consisting of aromatic lignin-type polymers of high molecular weight. For a complete removal of both types of material, Worrall found it necessary first to extract the bitumen with propanol in a Soxhlet extractor and, after removal of the excess propanol, to follow this by oxidation with 20 volume hydrogen peroxide. This procedure not only provides a clay specimen free from organic matter but by weighings at each stage affords a quantitative estimate of the total organic matter.

It appears therefore that the method often applied, i.e. simple oxidation with hydrogen peroxide, may not apply to all clays. Alternative methods, such as heat treatment at 300°C in pure oxygen, cannot be recommended because of the danger of partial decomposition of inorganic constituents at this temperature.

The IL/MA test

In its simplest form, this test, devised by Keeling (1961, 1966), consists of determining the ignition loss (%), IL, and the moisture absorption, MA, and hence calculating the ratio IL/MA. In this form, the test is rapid and easy to carry out. In principle, however, the test requires that the combined water, rather than the total ignition loss, be measured; this means that the latter has to be corrected for any organic matter or carbonates, etc. The moisture absorption MA is again basically very simple: it consists of placing pre-weighed, pre-dried samples of clay in a humidity chamber containing saturated NaCl solution which maintains the humidity at about 75%. After allowing time to equilibrate, the samples are removed, re-weighed and the increase in weight reported as MA (%). Whilst kaolinites and illites may reach equilibrium in 24 h, montmorillonites and other expanding clays may require up to one week. Nevertheless, the test has been found useful in classifying clays for the ceramic industry.

The principle of the test is essentially as follows. For a clay consisting of pure kaolinite, $Al_2Si_2O_5(OH)_4$, the combined water (corrected IL) is about 14·0%. The MA is essentially a measure of surface area, which for a well-crystallised kaolinite is relatively low; hence the ratio IL/MA is high for kaolinites. At the other extreme, the 'model' formula for a montmorillonite may be written as $Al_2Si_4O_{10}(OH)_2$, from which it is seen that the corrected IL is about 4·5%. Moreover, montmorillonites have a very high surface area, so that the MA value is now high, giving a relatively low IL/MA ratio. Disordered kaolinites and illites have intermediate IL/MA values. Typical figures are as follows:

Well-crystallised kaolinite	>7
Disordered kaolinite	2–3
Illite	0·7–1·0
Montmorillonite	<0·7

Where a clay is contaminated with non-clay minerals such as quartz, which contributes little to either ignition loss or moisture absorption, both are reduced in almost the same proportion and so their ratio is little affected.

X-RAY DIFFRACTION

X-rays are a form of electromagnetic radiation, produced when matter is bombarded by a stream of fast-moving electrons. The wavelength of X-rays is in general much smaller than that of visible light and these radiations therefore have great penetrative power, and are capable of ionising gases or blackening a photographic plate.

Just as visible light can be diffracted by a series of ruled lines on a glass plate (a diffraction grating), provided the distances between the lines are of the same order as the wavelength, X-rays can be diffracted by the atom-bearing planes of a crystal. If a beam of X-rays falls on a series of atom-bearing planes, each a distance d apart, at an angle θ, it follows that for a sharp diffracted beam to be produced:

$$n.\lambda = 2d.\sin\theta$$

where λ is the wavelength of the rays and n is an integer. This is because a sharp beam is only produced when diffracted rays reinforce, that is, when their path lengths differ by an exact multiple of the wavelength. The above expression is known as Bragg's Law, the integer being the 'order' of the diffraction.

Clays and in fact most ceramic materials are used in the form of a fine powder, consisting of a large number of very small crystals. A specimen in the form of a thin wire is made from a small quantity of the powder and is rotated inside a special camera containing a strip of photographic film. A narrow beam of X-rays of known wavelength, produced by bombardment of a copper or other suitable target, is directed on to the specimen. The diffracted beams emerge as a series of cones, which on striking the photographic film produce a series of short arcs. The orientation of the minute crystals is of course random and many produce no diffracted beam at all, but some will have planes in just the right orientation to fulfil the requirements of Bragg's Law. Each set of planes having separation distances d may produce a number of lines on the film for values of n from 1 to 3 or higher, but as a rule the reflections become weaker as the order increases. From the dimensions of the camera, and the positions of the axes, the value of θ can be obtained and from this, for first-order reflections, the value of d can be found. Values of d-spacings are often referred to a standard calcite crystal and are then expressed as kX units instead of Angstroms.* Several different sets of d-spacings may of course be present in any given crystal. The d-spacing for the $(0, 0, 1)$ planes is called the *basal spacing*, which is characteristic of the relevant mineral and may serve to identify it. In fact, the powder diffraction method is mainly a method of identifying minerals of known spacings, rather than elucidating their structures. A limited amount of information can be obtained in this way by comparing the spacings between various sets of planes, but for structural work a single-crystal X-ray photograph is necessary.

Instead of, or in addition to, a photographic film, a limited angle around the specimen may be scanned by a goniometer, attached to which is a detecting device capable of recording the relative intensities of diffracted rays.

Characteristics of the kaolin group

The kaolin group of minerals may be identified by the prominent basal reflections at 7·14 and 3·57 Å. For kaolinite, there are also strong reflections at 4·47, 4·36, 4·18, 2·49, 2·38, 2·34 and 2·29 Å. Whilst nacrite, dickite and meta-halloysite have similar basal spacings, other reflections differ sufficiently to enable them to be distinguished from kaolinite. It should be noted that hydrated halloysite, in contrast to the other members of the group, has a basal reflection at about 10·1 Å, falling to 7·2 Å on heating to 60°C. For disordered kaolinites, the peaks at 7·14 and 3·57 Å are broader

---

*1 kX unit = 1·002 02 Å; the Å unit may eventually be replaced by the nanometre (nm), an SI unit equal to 10^{-9} m.

and less intense, whilst peaks in the range 2·71 to 4·44 Å merge to a broad band as disorder increases (Fig. 61).

X-ray diffraction has been extensively used for the quantitative analysis of kaolinite and other minerals in natural clays. Although the height or area of the 7·14 Å basal peak does depend on the proportion of kaolinite in a mixture, the relationship is not linear because of self-absorption effects. One method that overcomes this complication is the use of an internal standard such as boehmite which is added in known amounts both to the

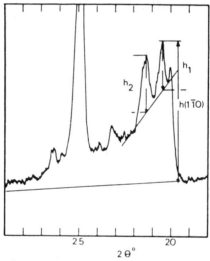

Fig. 61. Hinckley method for determination of crystallinity (from *Crystal Structures of Clay Minerals and their X-Ray Identification*, by kind permission of G. Brown, G. W. Brindley and The Mineralogical Society).

sample being analysed and to prepared standards of known kaolinite content. In all instances, the ratio of the peak height (or area) of the 7·14 Å kaolinite reflection to that of a suitable boehmite peak (usually 6·11 Å) is measured. This procedure not only provides a nearly linear calibration but effectively corrects for random instrumental variations. For further details, the reader should consult one of the reference works by Brown and Brindley (1980), Cullity (1978) or Klug and Alexander (1974).

In all quantitative work by X-ray diffraction, the choice of a suitable standard is crucial, as Brown and Brindley have pointed out. Ideally, the standard chosen should have the same degree of crystallinity as the specimen being analysed; if this is not practicable, appropriate corrections may be made for variations in crystallinity. Methods for measuring and correcting for crystallinity in the determination of kaolinite by X-ray have been described by Al-Khalissi and Worrall (1982).

The chlorite group

Minerals of the chlorite group are apt to be confused with those of the kaolin group, since many of the reflections are similar. However, the majority of chlorites show a basal reflection at about 14 Å, which is characteristic of this group. Where this reflection is weak or absent, auxiliary techniques may be used; for example, chlorite may be eliminated by treatment with dilute HCl, in which it is soluble, or kaolinite may be destroyed by heating to 600°C, leaving chlorite unaltered.

Montmorillonite minerals

The basal reflections of the montmorillonites vary with the hydration of the mineral, i.e. in the amount of inter-lattice water they contain, values from 10 to 15 Å having been reported. The introduction of certain organic polar liquids, e.g. glycerol, raises the basal spacing to a constant and characteristic value of 17·7 Å. This property of montmorillonite is most useful in its identification. The d-spacings around 2·5–2·64 Å are also characteristic of montmorillonites and are not affected by hydration. It may be difficult, however, to identify a particular member of the group by X-ray diffraction alone and for this a chemical analysis of the pure mineral is helpful.

The mica group

The mica group of minerals is well characterised by the basal spacing at about 10 Å. Although other spacings are similar for different micas, there is sufficient difference to distinguish, for instance, between muscovite and biotite. The so-called illites and hydrous micas show lines similar to muscovite but there is a marked weakening of certain lines.

In conclusion, it should be realised that the small particles and poor crystallinity of many clays often permit only the basal spacings to be identified, so that fine distinctions between the members of individual groups may not be apparent.

DIFFERENTIAL THERMAL ANALYSIS

Differential thermal analysis depends on the detection of the heat given out or absorbed when a phase change occurs in a substance which is being heated. In practice, the apparatus used consists of a refractory specimen holder, divided into two compartments; the material under test, in the form of a fine powder, is placed in one of these, and alongside it, in the adjoining compartment, is placed an inert material which does not undergo phase changes and which acts as a reference material. The specimen holder is provided with a set of thermocouples for measuring the temperature T of

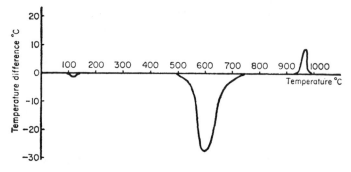

Fig. 62. Differential thermal analysis of kaolinite.

the inert material and also the *difference* in temperature between the inert and the specimen (ΔT). To achieve uniformity of temperature, the specimen holder is placed inside a cylindrical refractory block, which in turn fits closely inside an electrically heated tube furnace. The latter is controlled so that its temperature is raised uniformly from room temperature up to the desired maximum (usually 1000–1200°C) at a constant rate of 10°C/min. The thermocouple readings may be recorded automatically on a chart recorder.

Whilst no reactions are occurring, the temperature of the inert material will be the same as that of the specimen, but when a reaction involving a heat change occurs, the temperature of the specimen will deviate from that of the inert, according to whether the reaction is exothermic or endothermic. The thermocouples are so arranged that one set measures the temperature of the reference material (T) and the second set the temperature difference (ΔT). If therefore ΔT is plotted against T, a peak will occur whenever a phase change occurs. The 'zero line' on the graph is called the

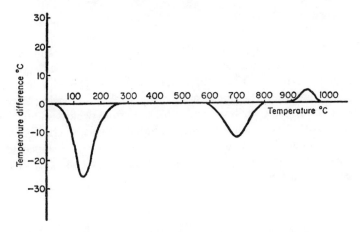

Fig. 63. Differential thermal analysis of montmorillonite.

base line and of course the specimen temperature tends to return to this line after a heat change by virtue of the dissipation of heat to the surroundings.

From the graph, it is clearly possible to determine the sign of the heat change (exothermic or endothermic), the approximate temperature at which it commences and, from the magnitude of the peak, the approximate proportion of reactive material. It should be appreciated that the temperature corresponding to the tip of the peak is somewhat arbitrary because it depends on the rate of heating; however, the standard rate of 10°C/min is widely accepted and peak temperatures based on this serve to identify many substances. Typical thermal analysis curves for well-crystallised kaolinite and for montmorillonite are shown in Figs 62 and 63 respectively.

Characteristics of kaolinite

As indicated in Fig. 62, a small endothermic peak occurs with kaolinite at about 100°C, caused by the evolution of adsorbed water. At about 500°C the main endothermic peak commences, the tip of the peak being at about 600°C, corresponding to the decomposition of the mineral, and the elimination of the hydroxyl groups as water. A further peak, which is exothermic, occurs at about 1000°C, and is associated with a recrystallisation process. This has been variously attributed to the formation of mullite, of alumina and of silicon spinel; the various hypotheses concerning this reaction were fully discussed in Chapter 8.

Disordered kaolinite

Sufficient differences are found between well-crystallised kaolinite and disordered kaolinite to warrant special attention. In disordered kaolinite, the initial peak due to adsorbed water is often larger than for well-crystallised kaolinite. The main endothermic peak is somewhat lower—usually at about 580°C or less—and the exothermic peak is smaller and broader. In fact, the exothermic peak is known to be very much affected by extraneous materials: for instance, it is suppressed by substances containing alkali metals; hence, by itself it is of less diagnostic significance than the main endothermic peak.

Halloysite

The main endothermic peak for halloysite occurs, as for kaolinite, at around 600°C, but for the hydrated variety there is in addition a subsidiary endotherm at 150°C. There is also the exothermic peak, as for kaolinite, at 1000°C.

Nacrite and dickite

The thermal analysis curves for nacrite and dickite closely resemble those of kaolinite. Dickite, however, appears to give a main endotherm at about 650°C—rather higher than the other kaolin minerals.

The montmorillonites

The montmorillonites all have in common a large endotherm at about 150°C, corresponding to the evolution of inter-layer water and water associated with the exchangeable cations. The peak varies in size and shape according to the nature of the cations and the lattice substitutions. The main endotherm, corresponding to the thermal decomposition of the mineral, occurs at about 700°C and there is a final exotherm at about 1000°C as for kaolinite.

Some authorities have reported multiple low-temperature peaks around 150 to 200°C for montmorillonites saturated with various cations, which they claim correspond partly to hydration of the surface and partly to hydration of the cation, but the interpretation of these multiple peaks is doubtful.

Chlorites

The chlorites show a wide variation in their thermal characteristics, due no doubt to the wide variety of possible compositions. In general, the main endothermic peak occurs between 500 and 700°C, with a second endothermic peak at about 800°C followed by an exothermic peak. The two endotherms are believed to be associated with the formation of water from hydroxyl groups in different environments, one associated with the mica-type units, the other with the brucite layers. The exothermic peak is believed to be caused by the recrystallisation of olivine from the decomposition products.

Micas

Although muscovite and related minerals are known to decompose at about 950°C, most normal micas show no thermal analysis peaks between room temperature and 1000°C. The illites and hydrous micas, however, do give endothermic peaks ranging from 500 to 700°C, but it is not certain whether these peaks are due to the mica itself or to an alteration product.

Accessory minerals

It is impossible to list all the possible accessory minerals likely to be associated with a naturally occurring clay, but the following are the commonest.

Quartz

Quartz shows a small peak at 573°C due to the α–β inversion, but this is so small that unless the apparatus is very sensitive it may not be detected.

Carbonates

The majority of carbonates decompose within the range of temperatures normally employed for thermal analysis. The evolution of carbon dioxide is associated with endothermic peaks at various temperatures; ferrous carbonate, for example, gives an endotherm at about 350°C, calcite at about 850°C.

Hydrated oxides

Hydrated oxides generally decompose, with the elimination of water, at a comparatively low temperature, from about 200–350°C. Typical of such oxides are gibbsite, $Al(OH)_3$, and goethite, $FeO(OH)$.

Pyrites

This mineral produces a low-temperature endotherm on decomposition, followed by an exotherm due to oxidation of ferrous iron to ferric.

Organic matter

A common impurity of clays is finely divided organic matter, resembling bituminous coal or brown coal. Such substances invariably give large exothermic peaks between 350 and 700°C, due to combustion, which may partly or wholly mask other peaks. For a successful thermal analysis on clays containing more than a few per cent of organic matter, it is necessary to remove the latter first by treatment with hydrogen peroxide.

THERMOGRAVIMETRIC ANALYSIS (TGA)

As in differential thermal analysis, this method employs an electrically heated furnace in which a specimen of clay or other material is heated at a constant rate (usually 5°C min^{-1}) from room temperature up to 1000°C or, in some circumstances, 1500°C. The specimen, contained in a small refractory crucible, is suspended from a sensitive recording balance, thus enabling simultaneous recordings of mass and temperature to be made. The earlier instruments consisted essentially of a two-pan analytical beam balance, of which one pan was connected to the specimen via a fused silica extension rod, the other being used as a counter-poise; the specimen weight was commonly 0·2 to 1·0 g. In the more recent instruments, an electronic micro-balance replaces the beam balance, enabling the sample weight to be reduced to as little as a few milligrams if required, depending on the anticipated weight loss. The furnaces are normally controlled by an electronic programmer which may be set to provide a range of heating rates.

Reactions that may be detected by weight loss include dehydroxylation of clay minerals, decomposition of carbonates, loss of combined water from hydrated oxides or from other hydrates such as gypsum, loss of sulphur from pyrites and sulphides and decomposition of organic matter. Occasionally weight gains may be recorded, e.g. oxidation of iron (II) to iron (III) oxide. Clearly, no information on phase changes, such as silica inversions, is obtained so that in some respects TGA is less informative than DTA. Like the latter, it may be used qualitatively or semi-quantitatively.

Of the clay minerals, kaolinite shows a weight loss commencing at just over 400°C (possibly lower for disordered varieties) due to dehydroxylation, extending to about 500°C; montmorillonites lose weight in two stages, commencing at 100–300°C, followed by a further stage between 300 and 850°C, depending on the nature of the mineral. The illites appear to lose water over a fairly broad range of temperature, from 400 to 700°C, again depending on the chemical nature and probably the degree of degradation. As with DTA, the steepness of the weight/time graphs depends to a large extent on the rate of heating.

INFRA-RED ABSORPTION SPECTROSCOPY

The absorption of infra-red radiation by matter arises from vibrations of atoms within a molecule and from rotations of the molecule itself. By observing the absorption of a substance over a range of wavelengths in the infra-red region (from about 1 to 100 μm), an absorption spectrum is obtained which is characteristic of the substance concerned. Hence, infra-red spectroscopy may serve to identify minerals in an unknown mixture; the same technique can also provide useful information on internal structure. Data on the nature of the —O—H— bond and of the environment of OH groups has been obtained from absorption measurements in the relevant region of the spectrum.

It has been customary among clay mineralogists to speak of *wave number* (i.e. the number of waves per cm) rather than wavelength. The wave number therefore is expressed as cm^{-1}, and although with the introduction of SI units this should now be m^{-1}, it is likely that the older system will persist for some time. It should be noted that, for easy conversion, wave number in cm = 10^4/wavelength in μm. The generally accepted range of infra-red radiations is therefore 100–10 000 cm^{-1}, whilst the region of interest for clays lies between 400 and 4000 cm^{-1}.

The absorption spectra of clay minerals in the infra-red region of the spectrum has been the subject of numerous investigations, among which may be mentioned those of Farmer (1971).

For kaolinite, there is a group of absorption peaks between 3500 and 3750 cm^{-1}, sometimes called the 'a-band', which is due to 'stretching' frequencies of the OH groups. The 'inner OH' groups, i.e., those completely enclosed by aluminium and oxygen atoms within the gibbsite layer, have a characteristic stretching frequency of 3620 cm^{-1}. The 'inner surface' OH groups, i.e., those at the boundary of the gibbsite layer and able to form hydrogen bonds with the adjoining silica layer, have characteristic stretching frequencies of 3697, 3669 and 3652 cm^{-1}. It is the latter three that are susceptible to disturbances of hydrogen bonding associated with ab displacements or isomorphous substitutions. For instance, an absorption peak at as high as 3700 cm^{-1} would indicate loss of hydrogen bonding, whilst merging of the 3669 and 3652 peaks into one band at 3658 cm^{-1}

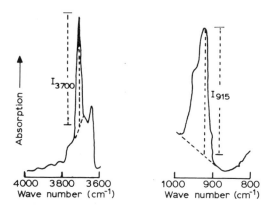

Fig. 64. Crystallinity by infra-red absorption.

indicates disturbance of the basal OH groups. Accordingly, the peak intensity at 3697, compared to that at 3620, has been suggested as an index of degree of crystallinity.

A second group of absorption peaks for kaolinite, sometimes called the c-band, occurs at much lower frequencies, namely at 910 and 938 cm^{-1}, these being attributed to Al—OH bending vibrations. There is also an Si—O— stretching vibration at 960 cm^{-1} (Fig. 64).

Montmorillonites give Al—OH and Si—O peaks in the same region as does kaolinite, but there is only one OH peak, namely that at 3620, which is consistent with there being only 'inner' hydroxyls in montmorillonite.

Dioctahedral micas show OH stretching frequencies at $3620\,\text{cm}^{-1}$, trioctahedral at $3700\,\text{cm}^{-1}$. The illite type of mica present in fireclay appears to have an OH peak at 3640 instead of 3620, however. Micas also exhibit Si—O—Al vibrations at 780 and $540\,\text{cm}^{-1}$ and an Al—OH bending vibration at $935\,\text{cm}^{-1}$.

Two general methods for preparing samples for infra-red work are in current use: the finely powdered material may be suspended in a liquid, the absorption peaks of which do not interfere with those of the sample; or a small quantity (approximately 1 mg) is ground with 0·2 g of potassium bromide and pressed into a disc in a 15 mm diameter die at a pressure of 20 tons. The latter method has been found most suitable for clays, but where absorptions are relatively weak, as with disordered kaolinites, it may be advisable to increase the ratio of clay to potassium bromide.

Infra-red spectroscopy has been successfully used by Toma and Goldberg (1972) for the determination of quartz but is of limited use for the determination of clay minerals because of spectral overlap and the effects of disorder on absorption intensities. The degree of crystallinity of kaolinite can, however, be assessed by this means, as demonstrated by Parker (1969) and by Neal and Worrall (1977).

OPTICAL MICROSCOPY

The limit of resolution of an optical microscope is determined partly by the wavelength of visible light and in practice partly by the maximum numerical aperture that can be obtained. If one considers two points in a microscopic field that are gradually brought closer together, they will appear to merge when they are still separated by a finite distance; this distance is known as the limit of resolution and is given by the expression $\lambda/2\text{n.a.}$, where λ is the wavelength of the light and n.a. the numerical aperture of the instrument. Since the numerical aperture can rarely exceed about 1·6 in practice, and the average wavelength of white light is $5\cdot3 \times 10^{-5}\,\text{cm}$, the theoretical limit of resolution is about $1\cdot7 \times 10^{-5}\,\text{cm}$ or $0\cdot17\,\mu\text{m}$. Since a high proportion of clay particles are near or below this value, only a limited amount of information can be expected from an optical study of the clay minerals.

Some of the larger particles of kaolinite, however, are recognisable under the microscope as regular hexagonal plates, whilst aggregates of montmorillonite may also be recognisable. In addition, it is often possible to measure the refractive index and birefringence of clay particles. The refractive index of the kaolin group is around 1·57, that for montmorillonite being about 1·50, higher values being obtained for iron-substituted varieties. Since the clay minerals are anisotropic, the figures quoted are mean values only. Optical methods have also been used to determine the orientation of clay particles in an electrostatic field. Again, when stained with suitable basic dyes, alteration products may be detected microscopically in some of the altered micas.

ELECTRON MICROSCOPY

In electron microscopy, a beam of electrons, focused magnetically, is employed instead of visible light; the electrons are scattered by solid objects such as clay particles, thus casting a 'shadow' on a fluorescent screen or photographic plate. In early work the clay particles were not treated in any way, but later methods were developed for coating the particles with a thin film of metal to improve the contrast and to show up the thickness of the particles. Another technique that has been used is the replica method, in which a freshly fractured surface of the clay is coated with cellulose nitrate, thus forming a replica of the particles. The replica is then coated with a thin film of aluminium or beryllium and the cellulose dissolved away. The resulting metal replica is then mounted in the instrument.

Because of the very short effective wavelength of the electrons (of the order of 1 Å), very high resolution and magnification are possible; for example, a magnification of 100 000 with a resolution down to 20 Å is quite common. Either positive or negative photographic prints may be prepared by electron microscopy, such prints being known as electron micrographs. Under favourable circumstances, they reveal the approximate size and shape of clay particles and sometimes their thickness, but no structural detail is discernible. Unfortunately, during the preparation of specimens, particles that were originally well dispersed frequently become reagglomerated, thus rendering estimation of size and shape very difficult. Moreover, unlike optical microscopy, electron microscopy does not permit any ancillary techniques such as polarisers and staining methods.

Nevertheless, much useful information has been obtained from the electron microscopy of clay minerals. It has been confirmed that kaolinite forms roughly hexagonal plates approximately 0·3–5 μm in diameter and 0·05–2 μm in thickness. In disordered kaolinites, particles as small as 0·02 μm have been reported. On average, the thickness of kaolinite particles is about one-eighth their diameter.

Dickite also exists as platy particles ranging in size from 2 to 8 μm and from 0·05 to 0·25 μm in thickness. The particles of nacrite appear to be less regular than those of kaolinite but with similar size ranges.

Halloysite differs from the other kaolin minerals in that it exists as tubular particles. A few such particles may be 1 or 2 μm in length, but the majority are less than 1 μm. It has been suggested that the tubular habit of halloysite is caused by the strain due to 'misfit' of the silicon–oxygen and aluminium–oxygen layers. The hydroxyl bonding and consequent greater thickness of the other kaolins presumably prevents tube formation.

In the montmorillonites, individual particles are difficult to discern, the general appearance being that of a fluffy, irregular mass of agglomerated particles. Where individual particles have been found, they appear to be extremely thin—approximately 0·002 μm in thickness—showing that they approach unit cell size. The average diameters are difficult to assess but are reported to be approximately 0·02–0·2 μm. It is becoming increasingly clear, however, that the apparent size of montmorillonite particles depends very much on the degree of dispersion achieved in their preparation; this is influenced, among other things, by the nature of the exchangeable cations.

Electron micrographs of illites resemble those of montmorillonite but the particles are somewhat larger and better-defined. Micaceous material found in fireclays has frequently been referred to as illite. Some relatively large particles of this material have been found (about 20–36 μm diameter), but much fine-grained material is also present that cannot be distinguished from kaolinite under the electron microscope, suggesting that composite particles of kaolinite and illite are present.

MÖSSBAUER SPECTROSCOPY

This relatively new instrumental method is worthy of mention since it has provided direct confirmation of substitution in kaolinite; namely, the replacement of Fe^{3+} for Al^{3+} in the octahedral layer.

Essentially, the method employs a radioactive source (for clays this is provided by a ^{57}Co source, which decays to ^{57}Fe, the nuclei of which are then in an excited state), which is moved mechanically at a velocity of a few millimetres per second towards a prepared specimen, which forms the 'target'.

The source emits γ-rays of definite frequency, which can be absorbed by the 'target' only if the latter contains Fe atoms, i.e. if resonance is achieved. The absorption is, however, modified by the Doppler Effect due

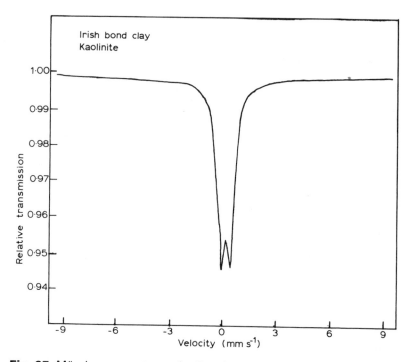

Fig. 65. Mössbauer spectrum of a disordered kaolinite (after removal of free iron oxide) (courtesy of C. Gomes).

to the relative motions of the source and the 'target', so that an absorption spectrum can be obtained by plotting source velocity against absorption (Fig. 65). This spectrum also depends on the environment of the Fe atoms, so that it is possible to distinguish, for instance, Fe^{3+} in the form of free iron oxide from Fe^{3+} in the kaolinite structure.

So far, only Fe^{3+} in octahedral sites has been found, there being no evidence of tetrahedral Fe. Since the method is only applicable to the heavier elements with a sufficiently long half-life, it cannot therefore be used to detect substituent elements such as Mg in octahedral sites or Al in tetrahedral sites.

REFERENCES

AL-KHALISSI, F. Q., and WORRALL, W. E., *Trans. J. Brit. Ceram. Soc.*, **81** (1982) 43.

ANDREASEN, A. H. M., and LUNDBERG, J. J. V., *Trans. Brit. Ceram. Soc.*, **29A** (1930) 239.
BROWN, G., and BRINDLEY, G. W. (eds), *Crystal Structures of Clay Minerals and their X-Ray Identification*, Mineralogical Society, 1980.
CULLITY, B. D., *Elements of X-Ray Diffraction*, Addison-Wesley, 1978.
FARMER, V. C., *Soil Sci.*, **112** (1971) 62.
HASHIMOTO, I., and JACKSON, M. L., *Clays and Clay Minerals*, 7th Nat. Cong. (1960) 102.
KEELING, P. S., *Trans. Brit. Ceram. Soc.*, **60** (1961) 217; **65** (1966) 463.
KLUG, H. P., and ALEXANDER, L. E., *X-Ray Diffraction Procedure for Polycrystalline and Amorphous Materials*, Wiley, London, 1974.
MEHRA, O. P., and JACKSON, M. L., *Clays and Clay Minerals*, 7th Nat. Cong. (1960) 317.
NEAL, M., and WORRALL, W. E., *Trans. Brit. Ceram. Soc.*, **76** (1977) 57.
PARKER, T. W., *Proc. Brit. Ceram. Soc.*, **13** (1969) 117.
TOMA, S. Z., and GOLDBERG, S. A., *Anal. Chem.*, **44** (1972) 431.
TROSTELL, L. J., and WYNNE, J. D., *J. Am. Ceram. Soc.*, **23** (1942) 1.
WORRALL, W. E., *Trans. Brit. Ceram. Soc.*, **55** (1956) 689.

READING LIST

H. BENNETT, and R. A. REED, *Chemical Methods of Silicate Analysis*, Academic Press and British Ceramic Research Association, 1971.
C. W. BUNN, *Chemical Crystallography*, Oxford University Press, 1946.
V. C. FARMER, *The Infra-Red Spectra of Minerals*, Mineralogical Society, London, 1974.
R. E. GRIM, *Clay Mineralogy*, McGraw-Hill, 1968.
W. F. HILLBRAND, G. E. F. LUNDELL, H. A. BRIGHT, and J. I. HOFFMAN, *Applied Inorganic Analysis*, Wiley, 1953.
G. F. KIRKBRIGHT, and M. SARGENT, *Atomic Absorption and Fluorescence Spectrometry*, Academic Press, 1974.
R. C. MACKENZIE, *The Differential Thermal Investigation of Clays*, Mineralogical Society, 1957.
C. R. N. STROUTS, J. H. GILFILLAN, and H. N. WILSON, *Analytical Chemistry*, Oxford University Press, 1955.
J. W. STUCKI, and W. L. BANWORT, *Advanced Chemical Methods for Soil and Clay Minerals Research*, D. Reidel Publishing Company, 1979.
J. ZUSSMAN, *Physical Methods in Determinative Mineralogy*, Academic Press, 1977.

Chapter 10

Refractory Raw Materials

ALUMINA

Structure

Aluminium oxide or alumina, Al_2O_3, exists in two principal forms, α- and γ-Al_2O_3; the so-called β-Al_2O_3 is not the pure oxide but an aluminate having the formula $Na_2O . 11Al_2O_3$. α-Al_2O_3, also known as corundum, is the commonest form of alumina and also the most stable; it is formed when any other type of alumina is heated. The table of co-ordination numbers (Table 4) shows that the radius of the Al^{3+} ion is somewhat greater than that of Si^{4+}, hence Al^{3+} normally has the higher co-ordination number of 6. In the corundum structure, each Al^{3+} ion is surrounded by six oxygen ions each oxygen being surrounded by four Al^{3+} ions, thus achieving electrical

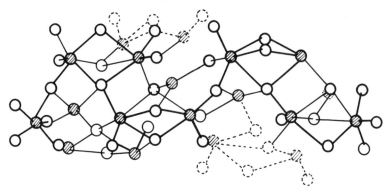

Key:— ○ Oxygen ◉ Al

Fig. 66. The structure of corundum.

neutrality with the net formula Al_2O_3. Some idea of the general appearance of the structure can be deduced from Fig. 66.

γ-Al_2O_3 has a structure based on that of the spinels, which are discussed in the section on chrome. A spinel has the general formula $A_3B_6O_{12}$, where A and B are di- and trivalent metallic ions respectively. In order to make γ-Al_2O_3 correspond to the above formula, the number of oxygen atoms must clearly be made up to 12, so we may write the formula as Al_8O_{12}. Comparing this with the spinel formula, we see that there are only eight cations instead of the normal nine, which is taken to mean that this is a *defect* structure, with one in every nine cations missing. With one 'vacant' site in every nine, the structure is still stable, though not as stable as corundum; it is converted into the latter when heated for a prolonged period at or above 1000°C.

Occurrence

Although aluminium in combination with oxygen and silicon as aluminosilicate is one of the most abundant constituents of the earth's crust, free alumina is comparatively rare. It is usually found in a hydrated form as bauxite rock, which consists of the minerals gibbsite, $Al(OH)_3$, diaspore and boehmite (both of formula $HAlO_2$). Bauxite is found in Jamaica, Guyana, Europe, Russia and elsewhere. Non-hydrated alumina, i.e. corundum, is a rare mineral, found in Greece and South Africa. Bauxite rock has been formed by the very severe weathering of various kinds of igneous rock under tropical conditions, in which circumstances not only the alkali oxides but also the silica was removed, leaving only hydrated oxides of aluminium and iron; this residue was consolidated to form bauxite rock. The bauxitic clays of Ayrshire are said to have originated in a similar way.

Commercial alumina is produced chiefly from bauxite, the total world production being millions of tons per year. The bauxite ore contains, in addition to aluminium oxides, ferric oxide and silica, which are removed by the *Bayer method*, as follows. First the ore is ground finely; it is then treated with sodium hydroxide solution in an iron autoclave under a pressure of 0·4 MPa and at 160–170°C. The alumina dissolves as sodium aluminate, as in the equation:

$$Al_2O_3 + 2NaOH = 2NaAlO_2 + H_2O$$
$$\text{sodium}$$
$$\text{aluminate}$$

The silica dissolves to form sodium silicate but the ferric oxide, being insoluble, remains as the so-called 'red mud', which is filtered off, thereby freeing the solution from iron. Carbon dioxide is then passed through the solution, decomposing the sodium aluminate to form aluminium hydroxide and sodium carbonate:

$$2NaAlO_2 + CO_2 + 3H_2O = Na_2CO_3 + \downarrow 2Al(OH)_3$$

The aluminium hydroxide, $Al(OH)_3$, is separated by filtration and washing, whilst the sodium silicate remains in solution and is thus removed. Finally, the aluminium hydroxide is calcined at 1000°C or higher, when it loses water of constitution, yielding alumina:

$$2Al(OH)_3 \xrightarrow{1000°C} Al_2O_3 + 3H_2O$$

It is desirable to calcine sufficiently so that α-Al_2O_3 rather than γ-Al_2O_3 is formed. Since bauxite ores are relatively rare, efforts are being made to extract alumina from clays. High-grade commercial alumina still contains 0·1–0·2% of Na_2O, 0·1% of CaO and traces of Fe_2O_3, TiO_2 and Cr_2O_3.

Physical properties

Natural corundum or α-Al_2O_3 is a water-white, very hard, crystalline substance. Gem stones such as sapphire and ruby consist chiefly of corundum, with traces of other oxides. Commercial alumina is a white powder, with a specific gravity of about 3·9, which, if calcined sufficiently, consists of minute crystals of α-Al_2O_3. It may be shaped by pressing, slipcasting or by electrodeposition; when the shapes are fired at 1700–1800°C they become extremely strong, even though the temperature of firing is well below the theoretical melting point (2050°C). This process is known as *sintering* and is employed in the manufacture of various pure-oxide ceramics. Articles of sintered alumina are hard, not readily attacked by acids or alkalis at high temperatures, and are capable of withstanding considerable changes of temperature without fracturing. Sintered alumina is therefore suitable for making crucibles, thermocouple sheaths and sparking plugs; its high electrical resistance makes it suitable for use in electrical insulation. Impure alumina in the form of raw bauxite is sometimes added to fireclay to increase its refractoriness.

Chemical properties

As already stated, α-Al_2O_3 is a very inert substance and resists most aqueous acids and alkalis. Fused caustic alkalis attack alumina slowly, forming aluminates, but for the analysis of alumina it is necessary to fuse with borax or with sodium peroxide to ensure rapid and complete decomposition. γ-Al_2O_3 is rather more reactive than α-Al_2O_3 and can be dissolved by hot, concentrated acids.

Articles are often made from α-Al_2O_3 by slipcasting, using deflocculated aqueous suspensions. The electrical charge on the surface of an α-Al_2O_3 particle is determined by the manner in which the amphoteric surface groups ionise. In aqueous suspensions, the surface $\geq$Al—O— groups are hydrated to —Al—O—H, which in acid solution releases OH^- ions, replaceable by Cl^- or other anions:

$$\geq Al\text{—}O\text{—}H \xrightarrow{\text{acid solution}} \geq Al^+ + OH^-$$

leaving a positive charge on the alumina surface. However, in alkaline solution H^+ ions are released, replaceable by other cations, forming a negative charge on the alumina surface:

$$\geq Al\text{—}O\text{—}H \xrightarrow{\text{alkaline solution}} \geq Al\text{—}O^- + H^+$$

Accordingly, alumina suspensions may be deflocculated by HCl at about pH 4·0, when Cl^- replaces OH^- as counter-ion; or by NaOH and other suitable bases, when the appropriate cation replaces H^+. Most other oxides, being more basic than Al_2O_3, ionise in water to produce positively charged particles and are therefore generally deflocculated by HCl. In some instances, better casts are obtained with suspensions in alcohol, using fatty acids as deflocculants.

Another forming process that makes use of the electrochemical properties of alumina is that of electrodeposition, which depends on the electrophoretic movement of the charged alumina particles to a charged electrode.

In very recent years there has been a marked revival of interest in wet-forming processes for producing oxide components used in the electronic and engineering industries. By applying the basic principles of colloidal chemistry as already outlined for clays, it has been found possible to prepare aqueous suspensions of various oxides suitable for slip-casting. The dispersing agents required to give the correct rheological properties may be simple acids or bases, as for alumina, but in some instances it may be preferable to use the more complex polymeric additives, e.g. polyacrylates, alginates or lignosulphonates. Components produced in this way have a more uniform microstructure than pressed components and moreover can be sintered to near theoretical density at lower temperatures. Although these wet processes may often be superior to dry methods, very precise control of particle size is necessary and it is in this area that much further research is needed.

REFRACTORY SILICATES

Apart from fireclays, which have a moderate refractoriness, a number of naturally occurring alumino-silicates having a somewhat higher refractoriness are also employed, as follows.

Sillimanite, kyanite and andalusite

These three minerals all have the empirical formula Al_2SiO_5 (or $Al_2O_3 \cdot SiO_2$) and contain some 63% of Al_2O_3. Sillimanite comes chiefly from South Africa and India, kyanite from the United States and India, andalusite from South Africa and the United States.

Structure

In all three minerals there are parallel chains of Al—O groups, linked sideways by silicon and aluminium ions alternately; the three structures are closely similar and differ only in detail, mainly in the co-ordination of the aluminium ions.

Properties

Sillimanite, kyanite and andalusite are non-plastic materials; refractory articles are therefore made from them by fine grinding, followed by a small percentage of water and a binder; the shapes are usually made by pressing, followed by firing to a high temperature.

When heated at about 1550°C all three minerals decompose to form *mullite* (the only stable compound of alumina and silica at high temperatures) and cristobalite:

$$3Al_2SiO_5 \xrightarrow{1550°C} Al_6Si_2O_{13} + SiO_2$$

Consequently, when refractory articles are made from any of the above three minerals, the final product is the same, since during firing the reaction shown takes place. The sillimanite group of minerals has a high refractoriness and is very resistant to attack by alkaline slags.

Mullite

Naturally occurring mullite is not very common; it is named after one of the few known deposits on the Isle of Mull, off the west coast of Scotland, and has the approximate formula $3Al_2O_3.2SiO_2$ or $Al_6Si_2O_{13}$, and so contains some 72% of alumina. Synthetic mullite is commonly used and can be made by heating a mixture of pure Al_2O_3 or bauxite with clay or sillimanite. Mullite is a common constituent of fired clay-based ceramics and under the microscope appears as long, prism-shaped crystals of approximately square cross-section.

Structure

X-ray studies have revealed that mullite has a crystal structure closely similar to that of sillimanite. However, it is difficult to reconcile this with the empirical formula, even if it is assumed, as some authorities, e.g. Rooksby and Partridge (1939) and Durovic (1963), have suggested, that aluminium substitutes for silicon in some of the tetrahedral sites in the sillimanite structure. Moreover, 'pseudo-mullites' have been reported which have a formula similar to that of sillimanite, whilst 'mullites' with an $Al_2O_3:SiO_2$ ratio as high as 2:1 have been found. It is therefore possible, as some have claimed, that a proportion of the alumina in these alumina-rich 'mullites' is in solid solution rather than part of the structure.

Properties

Mullite is very refractory and resists alkaline slags. At 1810°C it dissociates to form corundum and a siliceous liquid.

STEATITE

Steatite is mineralogically the same as talc, also known as French chalk; the term *steatite*, however, is usually reserved for the massive form.

Talc was formed by the hydration of magnesium-bearing rocks under pressure and may be derived from basic igneous rocks or from dolomite or marble. It occurs in many parts of the United States, in Germany, France, Morocco and Thailand.

Structure

Natural talc is not pure but contains aluminium, iron and calcium. As previously described (see section on the montmorillonites), talc is a hydrated aluminium silicate having the 'ideal' formula $Mg_3Si_4O_{10}(OH)_2$. Structurally, it is composed of a layer of brucite, $Mg(OH)_2$, interposed between two silica-type layers.

Properties

Despite its resemblance to the montmorillonites, talc has no clay-like properties; it has little or no plasticity or cation exchange properties, although it does have a pronounced cleavage along one plane, so that it is relatively soft and has a characteristic 'soapy' feel. In the massive block form it can be readily machined and fires to produce a strong body. When fabricated in powder form, clay is sometimes added as a binder.

When heated at about 900°C, talc is decomposed with the elimination of combined water. At still higher temperatures, around 1300°C, recombination occurs and clino-enstatite, $MgSiO_3$, is formed, with the elimination of silica:

$$Mg_3Si_4O_{10}(OH)_2 \xrightarrow{1300°C} 3MgSiO_3 + SiO_2 + H_2O$$

Uses of steatite

Steatite or talc, compounded with a little clay and some flux, is used in making *steatite bodies* for low-loss electrical insulation. With the addition of some 50% of clay, and some prefired material, talc is used for the manufacture of *cordierite bodies*. Their main constituent, the mineral cordierite, is a magnesium aluminium silicate of formula $2MgO \cdot 2Al_2O_3 \cdot 5SiO_2$; it is a good electrical insulator at high temperatures, with a very low thermal expansion coefficient and high thermal shock resistance. By replacing the feldspar of a conventional earthenware body by talc, a body of low moisture expansion, suitable for wall tiles, is obtained. For all these purposes, the talc should contain a minimum of iron oxide and alkalis.

MAGNESITE

Magnesite is one of the group of basic refractories that were introduced in 1880 to cope with problems encountered in steelmaking. Strictly speaking, the term *magnesite* stands for the mineral $MgCO_3$, which occurs in sedimentary rocks formed from the decomposition of ultrabasic igneous rocks containing olivine and other silicates of magnesium. It is often associated with limestones and dolomites, occurring in Austria, Czechoslovakia, Greece, Yugoslavia, Russia, Canada, the United States, Brazil and North-East China. There are only a very few minor deposits of magnesite in the United Kingdom, but *sea-water magnesite* (see below) has very largely replaced natural magnesite in the British ceramic industries.

Natural magnesite is always calcined before use as a refractory, when decomposition occurs, with the evolution of carbon dioxide:

$$MgCO_3 \longrightarrow MgO + CO_2$$

The product, magnesium oxide, is commonly referred to as 'magnesite', although it is no longer a carbonate. Just as quicklime readily slakes or combines with water to form the hydroxide, magnesium oxide tends, in the presence of atmospheric moisture, to combine with water to form $Mg(OH)_2$, which in refractory products causes a marked expansion and consequent failure. The tendency to hydration is markedly reduced, however, if the calcination temperature is 1500°C or higher, when the magnesia is said to be 'dead-burned'.

Sea-water magnesite

Sea-water magnesite is unique in that it is obtained by a chemical reaction, the basis of which is as follows. Sea water contains, among other things, sulphates and chlorides of sodium, potassium, calcium and magnesium. If calcium hydroxide is added to the sea water, the magnesium salts react to form sparingly soluble magnesium hydroxide, which is thus precipitated, with calcium chloride and sulphate as the chief soluble by-products. These, and any other soluble impurities, can then be filtered off leaving magnesium hydroxide as the residue. The calcium hydroxide required for the process is obtained by calcining dolomite, obtained in the United Kingdom from quarries at Thrislington, near Hartlepool; during calcination, carbon dioxide is evolved, the residue being a mixture of the oxides of calcium and magnesium. The residue is then slaked by the addition of water forming the respective hydroxides, the successive reactions being:

$$CaMg(CO_3)_2 \xrightarrow{heated} CaO + MgO \xrightarrow{H_2O} Ca(OH)_2 + Mg(OH)_2$$

Sea water is pumped through pipelines some 2000 ft (600 m) into the sea, and reacted with the treated dolomite in agitated reaction vessels. The reaction may be represented by the equation:

$$\begin{bmatrix} Mg(OH)_2 \\ Ca(OH)_2 \end{bmatrix} + \begin{bmatrix} MgCl_2 \\ MgSO_4 \end{bmatrix} \rightarrow 2Mg(OH)_2 + \begin{bmatrix} CaCl_2 \\ CaSO_4 \end{bmatrix}$$

'slaked dolomite'　　sea water　　magnesium hydroxide　　by-products

The precipitated magnesium hydroxide is then allowed to settle in large-diameter settling tanks, in which it sediments to form a thick sludge which is then pumped out, washed, dried and calcined at 1600°C. Since the sea water contains some calcium bicarbonate, which would precipitate the normal carbonate in the reaction vessels, a preliminary treatment with the requisite quantity of sulphuric acid is first carried out.

The present sea-water magnesia plant at Hartlepool was first built in 1938 and now produces some $250\,000 \times 10^3$ kg of magnesia per year (Fig. 67). It should be noted that the magnesia content of sea water is about

Fig. 67. A 100 m settling tank (photograph by courtesy of the Steetley Company Ltd).

0·2%, and even by enrichment with dolomite, some 300 kg of sea water needs to be processed to obtain 1 kg of magnesia. Some idea of the average composition of sea-water magnesite and of natural magnesite can be obtained from Tables 31a and 31b.

Structure of magnesium oxide

The crystal structure of magnesium oxide, or *periclase*, is illustrated in Fig. 68. This is one of the simplest structures and is very similar to that of rock salt, NaCl. Each magnesium atom is surrounded by six oxygen atoms; each

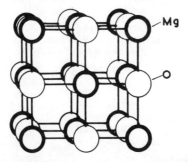

Fig. 68. The structure of magnesium oxide.

oxygen is in turn surrounded by six magnesium atoms, thus giving a 6:6 type of co-ordination. The entire lattice can be considered as built up from cubes, with oxygen and magnesium atoms at alternate corners. The periclase structure is the only known crystalline form of magnesium oxide.

Physical properties

Pure magnesium oxide is a white powder, having a specific gravity of 3·65 when calcined and a melting point of 2800°C. The refractoriness of commercial magnesite bricks is considerably lower than that of the pure oxide, since the strength of the brick depends on secondary crystalline products which are formed during firing. Magnesite is used extensively as a lining for the hearth in basic open-hearth steel furnaces.

Chemical properties

The chief value of calcined magnesite is its resistance to attack by basic slags containing iron oxide and calcium oxide. On the other hand, it is

Table 31(a)
Composition of a Sea-water Magnesite
(By courtesy of the Steetley Company Ltd)

Oxide	Weight (%)
SiO_2	0·8
Al_2O_3	0·5
Fe_2O_3	1·3
CaO	0·8
B_2O_3	0·15
MgO	96·4

Table 31(b)
Composition of a Typical Natural Magnesite
(By courtesy of C. W. Hardy and the *Refractories Journal*)

Oxide	Weight (%)
SiO_2	1·50
Al_2O_3	0·07
Fe_2O_3	0·54
CaO	2·90
B_2O_3	0·01
Cr_2O_3	0·01
MgO	94·97

susceptible to attack by acidic oxides such as silica, combining to form low-melting compounds. Even when calcined, magnesite is attacked by strong acids; this property is made use of in the analysis of magnesite.

DOLOMITE

The mineral dolomite is the double carbonate of calcium and magnesium, having the formula $CaMg(CO_3)_2$. It occurs in the Permian Magnesian Limestone and in the Carboniferous Limestone (see Table 8). The Permian deposit runs along the eastern side of the Pennines from Nottingham in the south to Coxhoe and Hartlepool in the north. Deposits of dolomite are also found in South Wales. Dolomite rocks have been formed by a number of complex reactions, two of which may be mentioned: (1) the action of magnesium-containing solutions on limestone; (2)

the precipitation of magnesium carbonate from supersaturated solutions. The British dolomites have a CaO:MgO ratio corresponding closely to that of the ideal formula, but traces of silica, iron oxide, etc., are present as impurities. Considerable variations in CaO:MgO ratio are quite possible, however. Over 10^9 kg of the raw material are processed annually for the construction and maintenance of the linings of steel-melting furnaces. It has found extensive use in the more recent BOS processes. The preparation of dolomite for use as a refractory is similar to that for magnesite; the raw material is calcined, a mixture of calcium and magnesium oxides being formed, with the liberation of carbon dioxide.

Structure

The structure of calcined dolomite is essentially that of a closely associated mixture of calcium oxide and magnesium oxide, both of which have the rock salt or periclase structure shown in Fig. 68.

Properties

The calcined form of dolomite is normally available as a porous, granular white powder of bulk density $2 \cdot 6$–$3 \cdot 0$ g ml^{-1}. The principal disadvantage of calcined dolomite is its tendency to hydrate, causing bricks made from it to crumble. This hydration tendency is due largely to the high proportion of lime and persists even after calcination at 1600°C. To overcome this, dolomite can be 'stabilised' by the addition of a silicate such as talc, which combines with the free lime to form calcium silicate. Alternatively, the dolomite bricks may be bonded with tar, thus protecting them from atmospheric moisture and affording a storage life of a few months. Dolomite is slightly more susceptible to attack by iron-rich slags than is magnesite but has nevertheless proved very effective for linings in the newer steelmaking processes using oxygen.

CHROME

Chrome ore is a basic rock containing chromium, iron, aluminium, magnesium and oxygen combined together as a complex crystalline compound known as a *spinel*. In addition, the mineral serpentine, $Mg_3Si_2O_5(OH)_4$, is sometimes present as a gangue mineral (one that is not of economic value), an impurity.

Structure

Spinels have the general formula AB_2O_4, where A stands for a divalent cation (e.g. Fe^{2+}, Mg^{2+}, Cu^{2+}, Mn^{2+}), and B for a trivalent cation (e.g. Cr^{3+}, Fe^{3+}, Al^{3+}). The minerals chromite, $FeCr_2O_4$, magnetite, $Fe^{2+}Fe_2^{3+}O_4$, and magnesium spinel, $MgAl_2O_4$, all belong to this group. In a *normal spinel* (Fig. 69) the atoms are so packed that each A atom is co-ordinated with *four* oxygen atoms, i.e. in tetrahedral co-ordination, whilst each B atom is co-ordinated with *six* oxygen atoms, i.e. in octahedral co-ordination. Indicating the respective oxygen co-ordinations by Roman numerals, the formula of a *normal spinel* can therefore be written as:

$$A^{IV}B_2^{VI}O_4$$

There is, however, another type, known as an *inversed spinel*, in which the tetrahedral positions are occupied by B atoms, whilst the A atoms and the remainder of the B are randomly distributed throughout the octahedral positions. Thus, the formula of an *inversed spinel* may be written as:

$$B^{IV}(AB)^{VI}O_4$$

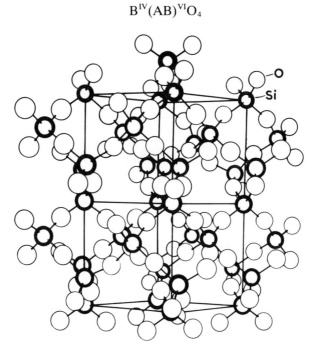

Fig. 69. The structure of normal spinel.

Ferrous chromite, $Fe^{IV}Cr_2^{VI}O_4$, is a normal spinel; magnetite, $Fe^{IV}Fe_2^{VI}O_4$, is an inversed spinel, one of the octahedral Fe atoms being divalent, the other trivalent. In chrome ore, the principal constituent is a mixed spinel in which the 'A' positions are occupied by Fe^{2+} and Mg^{2+}, and the 'B' positions by Cr^{3+}, Al^{3+} and Fe^{3+}.

Unlike magnesite, chrome ore retains the same structure (spinel) when heated, although some of the ferrous iron is oxidised; at the same time, the siliceous impurities are decomposed and afterwards recombine to form new products. The composition of a typical chrome refractory is shown in Table 32.

Occurrence and uses of chrome ore

Chrome ore is found in Russia, the Philippines, North America, Zimbabwe, Turkey and Greece, and is usually extracted by quarrying. Chrome refractories were first introduced in about 1880 as a 'neutral' zone between the acid and basic courses of steelmaking furnaces. Nowadays, chrome ore is usually mixed with 'dead-burned' magnesite for the production of *chrome-magnesite*, an important refractory which has made the 'all-basic' open-hearth steel furnace an economic possibility.

Physical properties

Chrome ore is a reddish-brown rock, with a specific gravity of 4·3–4·6, depending on composition. Being a mixture, it has no well-defined melting point, but is said to soften over about 1700°C. Under a load of 0·34 MN m^{-2} (50 lb in^{-2}), however, it fails completely between 1370 and 1470°C.

Chemical properties

Chrome ore is a comparatively inert substance. It is but little attacked by basic oxides or acidic oxides at high temperatures and for this reason it can be used as a neutral zone in furnaces containing both acid and basic refractories. The ore can be dissolved by the action of hot, concentrated sulphuric acid, which property is made use of in the analysis of chrome ores. It is, however, only incompletely dissolved by the other common acids and resists the action of caustic alkalis.

A phenomenon peculiar to chrome refractories is that a marked expansion, causing failure, occurs when they are in contact with iron oxide slags. Detailed examination of this reaction has shown that it is associated with the magnetite in the slag diffusing into the chrome spinel, which then causes an expansion. It is significant that similar effects occur in general when a normal spinel reacts with an inversed spinel, the result being a solid solution of the two components. The total expansion cannot all be accounted for by this chemical effect, however, and it is likely that disturbance of the original dense packing also occurs.

Table 32
Composition of a Typical Chrome Refractory
(Analysis by courtesy of Consett Refractories)

Oxide	Weight (%)
SiO_2	4·0
Al_2O_3	28·0
Fe_2O_3	16·3
Cr_2O_3	33·0
CaO	0·3
MgO	17·5

RUTILE

Rutile, a form of titanium dioxide, TiO_2, is a mineral associated with sands and sandstones; it is found in Travancore (India) and Australia. It is relatively inert, has a specific gravity of 4·2, and is insoluble in water and acids, except sulphuric acid, but is soluble in caustic alkalis.

As for alumina, the amphoteric nature of titanium dioxide results in aqueous suspensions acquiring a positive charge in acidic solutions and a negative charge in alkaline solutions, with an isoelectric point at about pH 5·0. Satisfactory casts from titanium dioxide are however best formed from alkaline suspensions of about pH 10.

The chief interest in rutile as a ceramic material lies in its high relative permittivity (about 110), which makes it a valuable material for electrical condensers. With certain other oxides, e.g. CaO and BaO, it combines to form titanates of general formula $MTiO_3$, which have relative permittivities of several thousand or more; the compositions of the titanates can be adjusted to produce a series of dielectrics having a wide range of properties.

ZIRCONIA AND ZIRCON

Zirconia occurs as the minerals zircite and baddeleyite, in Brazil and Sri Lanka. Pure zirconia has a specific gravity of 6·3 and is insoluble in water but soluble in sulphuric and hydrofluoric acids. It has a very high melting point, approximately 2700°C, and can therefore be used as a refractory material, although cost prevents its use on a large scale. Like silica, zirconia undergoes an inversion, accompanied by a volume change, at 1000°C, but may be stabilised by the addition of 10–20% of CaO.

Zirconium silicate or zircon, $ZrSiO_4$, occurs in India, Madagascar and Australia. It is a refractory material and has also been used in the manufacture of zircon porcelain for electrical insulation.

REFERENCES

DUROVIC, S., *Ber. deut. keram. Ges.*, **40** (1963) 287.
ROOKSBY, H. P., and PARTRIDGE, J. H., *J. Soc. Glass Tech.*, 100, **23** (1939) 338.

READING LIST

J. H. CHESTERS, *Steelplant Refractories*, United Steel Companies, 1973.
F. H. NORTON, *Refractories*, McGraw-Hill, 1968.
F. and S. SINGER, *Industrial Ceramics*, Chapman & Hall, 1963.

Chapter 11

Miscellaneous Raw Materials

FLUXES

A flux is a substance that is added to a body to enable it to fuse more readily. In ceramics, and in the pottery industry in particular, fluxes are incorporated in the body in order to lower the temperature at which liquid forms during firing. This liquid, when cooled, forms a glass, which binds the grains of the body together. By means of fluxes, strong articles of pottery or porcelain can be produced by being fired at 1100–1200°C, a temperature much lower than could be used in the absence of a flux. It is not desirable to lower the softening point further, since the higher proportion of liquid formed would then cause deformation and loss of shape.

For siliceous materials the most effective fluxes are materials rich in alkali oxides—Na_2O, K_2O or Li_2O. Calcium and magnesium oxides also act as fluxes, normally in combined form. Cost and availability are the main factors that influence the choice of materials; hence, naturally occurring minerals that contain the above oxides are preferred to pure chemicals.

SODA- AND POTASH-BEARING MINERALS

The more important of these minerals are compounds of Na_2O or K_2O with silica and alumina.

The feldspars

The feldspars are a group of minerals that all have similar chemical formulae. For the ceramic industry the most important are:

Potash feldspar (orthoclase) $KAlSi_3O_8$
Soda feldspar (albite) $NaAlSi_3O_8$
Lime feldspar (anorthite) $CaAl_2Si_2O_8$

Each may be considered to consist of four unit formulae of silica, Si_4O_8, in which one or two of the four silicon atoms has been replaced by Al and the charge deficiency made up by the addition of one atom of sodium, potassium or calcium.

The arrangement of the atoms differs from that in silica minerals, however. The silicon and oxygen atoms are linked so as to form four-membered rings, each ring containing four oxygen atoms; in orthoclase and albite the rings also contain three silicon atoms and one aluminium atom, whilst in anorthite they contain two silicon and two aluminium

Table 33
Composition of a Norwegian Feldspar

Oxide	Weight (%)
SiO_2	64·3
Al_2O_3	20·0
Fe_2O_3	0·18
TiO_2	0·1
CaO	1·5
MgO	0·1
Na_2O	4·9
K_2O	8·4
Li_2O	0·1
Loss on ignition	0·5

atoms. These four-membered rings are linked with similar rings to form chains, which in turn are cross-linked with similar chains via Si—O—Si groups, forming a three-dimensional framework. The sodium, potassium and calcium ions, being relatively large, are situated in the cavities within the framework. Part of the structure of a feldspar chain is shown in Fig. 22, where the larger circles represent oxygen, the small black circles silicon. The silicon and aluminium atoms have their normal fourfold co-ordination with oxygen, whilst the larger sodium, potassium and calcium ions are in eightfold co-ordination.

The chemical composition of a typical naturally occurring feldspar is shown in Table 33. The low iron oxide content indicates that it is suitable for use as a flux in whitewares. Of the fluxes used in the ceramic industry, the feldspars are the most important but one or two other minerals are used and are worthy of mention.

Nepheline

Nepheline, $Na(AlSi)O_4$, has a structure very similar to that of tridymite, the positions of the silicon and oxygen atoms corresponding closely. Nepheline may in fact be regarded as two unit formulae of tridymite, Si_2O_4, in which one of the silicon atoms is replaced by aluminium; the charge deficiency is then made up by the addition of sodium ions, which fit into the large voids in the tridymite framework. Like tridymite, nepheline converts to a cristobalite type of structure at high temperatures.

Lepidolite

Lepidolite is a lithium-bearing mineral, the constitution of which is represented by the idealised formula $K(AlLi_2)Si_4O_{10}(OH)_2$. It is a mica in which substitution of two lithium atoms for aluminium in the octahedral layer is balanced by potassium. The lithium-bearing minerals *petalite*, $LiAlSi_4O_{10}$, and *spodumene*, $LiAl(SiO_3)_2$, are also used as fluxes.

Wollastonite

Wollastonite, $CaSiO_3$, has also been used as a flux in bodies to obtain a low moisture expansion. This mineral has a ring structure.

Occurrence of alkali-bearing minerals

Deposits of relatively pure feldspar are not very common, being confined principally to Norway, Sweden, Russia and the United States. Rocks containing feldspar and other alkali-bearing minerals are much more abundant, such as the granites and pegmatites, which contain feldspar, mica and quartz in close association. Pegmatites are occasionally used as a flux, but the granites are generally too impure for this purpose and are often difficult to extract. There are considerable potential reserves of granite in the south-west and north-west of England and in Scotland, but these have not so far been exploited by the ceramic industry.

An altered granite known as china stone or Cornish stone, occurring in the St Austell granite of Cornwall, has been used extensively as a flux,

Fig. 70. A typical Cornish stone quarry near St Austell (photograph by courtesy of English Clays, Lovering, Pochin & Co. Ltd).

although it has in recent years been replaced to a considerable extent by imported feldspars. China stone is an intermediate product in the alteration of granite by hot, acidic gases to form china clay, as previously described; its chief virtue as a flux lies in its low iron content, the only troublesome impurity being fluorine-bearing minerals, notably fluorite, CaF_2, derived from the hydrothermal action; this latter impurity may be removed by the process of froth flotation. Quartz and mica are present in addition to feldspars but the former substances do not adversely affect the value of the stone as a flux. Cornish stone is extracted by blasting and quarrying; a stone quarry is depicted in Fig. 70.

Nepheline syenite

Nepheline syenite is a rock containing the mineral *nepheline* $Na(AlSi)O_4$ (in addition to sodic or potassic feldspars plus accessory minerals), the structure of which has already been discussed. This rock, which is found chiefly in the United States and Scandinavia, is refined to some extent before being put on the market. The chemical compositions of some typical Norwegian nepheline syenites are shown in Table 34. The high alkali ($Na_2O + K_2O$) and the low iron oxide contents should be noted.

Table 34
Compositions of Some Norwegian Nepheline Syenites (%)

Sample	Na_2O	K_2O	Fe_2O_3	SiO_2	Al_2O_3	Loss
1	6·87	7·07	1·20	52·3	—	2·60
2	7·36	8·80	2·10	53·01	—	1·60
3	7·42	8·20	2·32	53·6	—	1·02
4	7·90	9·08	2·25	52·7	—	1·87
5	8·14	8·75	2·13	52·6	—	1·79
6	7·36	8·80	1·68	53·0	—	1·21

Physical and chemical properties

Granites and pegmatites are extremely hard, grained rocks that are difficult to crush and grind. They are generally light in colour, the shade depending on the chemical composition, e.g. an iron-rich granite is reddish-brown or pink.

Cornish stone is somewhat softer than granite, depending on the degree of geological alteration. It contains very little iron oxide and is therefore an off-white colour; under the microscope, purplish flecks of fluorite may be seen in certain varieties. There are four main types of Cornish stone, differing in degree of alteration, these being (a) hard purple, (b) mild purple, (c) hard white and (d) soft white. The 'purple' varieties are richer in alkali than the 'white' varieties and are therefore the more powerful fluxes. The fluorite content decreases in the order given above. In practice, several different varieties are blended in order to produce a body having a suitable vitrification range. The chemical compositions of some selected samples of Cornish stone are shown in Table 35.

BONE

Strictly speaking, bone functions mainly as a 'filler' in bone china bodies and acts only to a limited extent as a flux. The main source of bone for the industry is that from cattle, because of its low iron content. It is prepared by treating the bone with steam to remove fat and gelatin and is then calcined at 800–1000°C. The product, bone ash, is then wet-ground to reduce 75–80% of the material to less than 10 μm and allowed to age for 3–4 weeks, the latter process imparting some degree of plasticity—a useful property in china bodies, which otherwise derive their plasticity almost entirely from moderately plastic china clay. It is likely that some of the plasticity of bone is due to residual organic matter, but with the present tendency to use higher calcination temperatures, the proportion that remains is inevitably less than formerly. A further consequence of the calcination is that 'slop' bone (aqueous suspensions) is appreciably alkaline, presumably because of free lime.

Mineralogical composition and structure of bone

Raw bone is essentially hydroxyapatite, $Ca_5(OH)(PO_4)_3$, with traces of iron and calcium carbonate. The chemical composition appears to deviate somewhat from the 'ideal' formula, there being rather less hydroxyl than the formula requires. When calcined, there appears to be little decomposition of hydroxyapatite, but the calcium carbonate is presumably decomposed to the oxide.

Firing reactions of bone

Earlier work, by St Pierre (1955), on the constitution of fired bone china showed that the main phases present in the final product were: anorthite, $CaAl_2Si_2O_8$; tricalcium phosphate, $Ca_3(PO_4)_2$; and a complex glass, which formed the bond. Investigations by Beech and associates indicated that during firing the whole of the metakaolin formed by decomposition of the

Table 35
Chemical Analyses of Some Varieties of Cornish Stone

Sample no.		SiO_2	Al_2O_3	Fe_2O_3	TiO_2	CaO	MgO	Na_2O	K_2O	Li_2O	Loss	F
Hard purple	1	71·9	15·4	0·2	tr.	1·8	0·4	4·1	4·2	tr.	1·4	1·0
	2	71·8	15·0	0·2	0·2	1·9	0·2	3·3	4·6		1·6	1·0
	3	71·6	15·8	0·1	0·2	2·4	0·1	3·6	3·7		1·7	1·2
	4	71·6	15·8	0·2	0·2	2·1	0·1	3·2	4·3		1·9	1·3
Mild purple	1	72·3	15·4	0·3	0·2	1·7	0·2	3·4	4·1		1·5	0·9
	2	72·9	15·5	0·4	0·1	0·8	tr.	0·9	6·9		1·7	0·3
	3	72·3	16·1	0·1	0·1	1·8	0·1	5·2	2·1		1·8	0·8
	4	71·9	16·0	0·2	0·1	2·1	0·2	3·1	4·1		1·8	1·5
	5	72·3	15·4	0·2	0·2	2·1	0·2	2·9	3·9		2·1	1·1
Hard white	1	73·1	16·8	0·2	0·2	0·9		1·2	4·7		2·5	0·3
Soft white		72·8	17·0	0·2	0·2	0·9		2·7	3·8		2·3	0·7

china clay reacted with the stoichiometric proportion of calcium to form anorthite, so that the reaction may be represented by the equations:

$$Al_2Si_2O_5(OH)_4 \longrightarrow Al_2Si_2O_7 + 2H_2O$$
$$\text{kaolinite} \qquad \text{metakaolin} \quad \text{water}$$

$$Al_2Si_2O_7 + CaO \longrightarrow CaAl_2Si_2O_8$$
$$\text{metakaolin} \qquad\qquad \text{anorthite}$$

It would appear that the remainder of the lime, or the greater part of it, is converted to tricalcium phosphate; it is also possible that some CaO, together with the soda and potash in the feldspars, forms a liquid which, on cooling, produces a complex glass.

PLASTER OF PARIS

Plaster of Paris, or calcium sulphate hemihydrate, $CaSO_4 \cdot \frac{1}{2}H_2O$, is prepared by heating the dihydrate, or gypsum, $CaSO_4 \cdot 2H_2O$, at 120–160°C:

$$CaSO_4 \cdot 2H_2O \xrightarrow{120-160°C} CaSO_4 \cdot \tfrac{1}{2}H_2O + 1\tfrac{1}{2}H_2O$$
$$\text{gypsum} \qquad\qquad \text{plaster of Paris}$$

Gypsum is a widely occurring mineral, believed to have been formed by the evaporation of sea water. In the United Kingdom, the deposits are of Jurassic, Triassic and Permian age. Gypsum is found as *alabaster*, a fine-grained rock; or as transparent crystals, when it is known as *selenite*.

On first heating gypsum the temperature rises rapidly to about 128°C, when the mixture appears to boil due to rapid evolution of water. The temperature remains at around 128°C until the first rapid reaction has ceased; this stage is called the 'first settle'. On further heating a second shorter period of 'boiling' commences, followed by a second quiescent period called the 'second settle'. At this stage the combined water content is probably below $\frac{1}{2}H_2O$, and the plaster contains some γ-anhydrite, formed by further dehydration:

$$CaSO_4 \cdot \tfrac{1}{2}H_2O \xrightarrow{160°C} \gamma\text{-}CaSO_4 + \tfrac{1}{2}H_2O$$
$$\text{plaster} \qquad\qquad \text{anhydrite}$$
$$\text{of Paris}$$

Heating above 163°C and up to 800°C results in conversion of γ-anhydrite to β-anhydrite or 'dead-burned' plaster. This commercial process yields mainly the so-called β-plaster; a harder-setting, fine-grained product known as α-plaster can be obtained by autoclaving gypsum in steam at 120–130°C or by boiling in 30% aqueous calcium chloride.

Setting of plaster

Both the hemihydrate and γ-anhydrite can rehydrate in the presence of water, forming gypsum; at the same time, provided the powdered material is thoroughly mixed with an excess of water to a homogeneous paste, the mixture sets to a hard mass of considerable strength. Although the prime mechanism of setting is the chemical rehydration of the plaster, the set product owes its strength to the interlocking of the long needle-like crystals of gypsum. For mould-making, a typical plaster–water mixture would contain 78–90 parts by weight of water to 100 parts of plaster. The setting of plaster is a strongly exothermic reaction. During the later stages of setting, the plaster undergoes a small but significant expansion of the order of 0·2–0·3% linear. This cannot be accounted for by the chemical reaction, since the density of the set material, i.e. gypsum, is very slightly *greater* than that of the hemihydrate; it must therefore be due to an edging apart of the crystals as setting occurs. As with all bodies formed by the compaction of a powder, set plaster is porous, air occupying the voids left as the water evaporates during the latter stages of setting. It is this pore structure which enables the plaster to be used as a mould because the small channels connecting the pores take up water by capillary action.

Unlike the other forms, β-anhydrite rehydrates very slowly and has no setting properties.

Control of properties

The setting rate of untreated plaster of Paris depends on the proportion of hemihydrate, β- and γ-anhydrites, impurities derived from the rock as mined, and on the particle size of the product. These factors are difficult to control, with the result that successive batches of plaster vary considerably in setting time, setting expansion and set strength. Fortunately, both setting time and expansion can be modified considerably by the addition of suitable electrolytes. Among those that accelerate the setting are potassium sulphate and Rochelle salt, in addition to the majority of sulphates and chlorides of the alkali metals. Retarders include borax, acetates and citrates, gelatin, starch and other organic colloids. It is found that both accelerators and retarders reduce setting expansion very

Table 36
Addition of Electrolytes to Plaster

Plaster/water ratio	Borax (%)	K_2SO_4 (%)	Setting time (min)	Linear expansion (%)
100:60	0·0	0·0	10·0	0·29
100:60	0·0	1·0	3·5	0·15
100:60	0·0	4·0	1·0	0·05
100:60	0·4	0·0	13·0	—
100:60	1·0	0·0	37·0	—
100:60	0·4	1·0	5·0	0·05
100:60	0·4	4·0	3·0	0·05
100:60	1·0	1·0	19·0	—
100:60	1·0	4·0	9·0	—

markedly; this property may be useful where plaster casts are made to a specified size. Where the main object is reduction of setting expansion, the effect of an accelerator may be balanced against that of a retarder. Some typical results of adding various amounts of potassium sulphate and borax to a plaster are given in Table 36. The percentages of electrolyte quoted refer to the strength of the solution (weight/volume basis) used in place of pure water. Where no value is quoted, the setting expansion was so small that it could not be measured. It will be clear that a wide range of setting times and expansions can be produced by suitable addition of potassium sulphate, borax or both, the effect of the two in admixture being roughly additive.

It is important during such measurements to maintain a constant ratio of plaster to water, since the latter also has a marked effect on setting time, expansion, strength and porosity. The effect of the plaster–water ratio on setting time and compressive strength is shown in Table 37. The plaster–water ratio very largely controls the porosity of set plaster; the higher the ratio, the less porous is the final set product. Control of the plaster–water ratio is therefore important in the making of moulds for the pottery industry.

Owing to the readiness with which gypsum dehydrates, care has to be taken in drying moulds that have been used for slip-casting, and a drying temperature not higher than 40°C is usually recommended. Plaster moulds may also be attacked by deflocculants used in casting slips: since the same

Table 37
Effect of Plaster–Water Ratio on Setting Time
and Compressive Strength of Plaster

Plaster–water ratio	Setting time (min)	Compressive strength $(MN\,m^{-2})$	
		2 hr	48 hr
100:80	10·5	4·63	8·50
100:60	7·25	7·73	13·13
100:45	3·25	10·82	26·26

mould is used repeatedly, the effect is cumulative and frequent cleansing may be necessary.

Thermal analysis studies

Detailed differential thermal analysis studies of various plasters and gypsums have been carried out by Holdridge. Gypsum gives a double endothermic peak, the first portion between 130 and 160°C, the second at 203°C. Comparison of this with the thermal curve of plaster, which gives a single endothermic peak at 203°C, corresponding to the conversion of hemihydrate to γ-$CaSO_4$, indicates that the first portion must be associated with the primary dehydration of gypsum to hemihydrate. Following the endothermic peaks, an exothermic peak given by both gypsum and plaster of Paris occurs at 220–360°C, due to, conversion of γ-anhydrite to β-anhydrite. An interesting feature of this exotherm is that whereas with β-plaster the peak occurs at 360°C and is well separated from the previous endothermic peak, for α-plaster the exotherm commences at 220°C, immediately following the second endothermic peak, and is very sharp. This difference has been attributed to a greater degree of disorder in the α-plaster structure and affords a useful method of distinguishing between the two varieties. It should be borne in mind, however, that quite frequently both varieties may be present, as in some plasters produced by open-hearth processes.

Extraction of gypsum

Gypsum suitable for the manufacture of plaster of Paris is found in considerable quantity in the United Kingdom. It is worked near Penrith, in other parts of Cumbria and Yorkshire, between Uttoxeter and Newark in the Midlands, and at Robertsbridge in Sussex. Some deposits are worked by quarrying, but underground mining is used for most deposits.

REFERENCE

ST PIERRE, P. D. S., *J. Am. Ceram. Soc.*, **38** (1955) 217.

READING LIST

D. G. BEECH, *The A.T. Green Book*, British Ceramic Research Association, 1959, p. 49.
S. DUROVIC, *Ber. deut. keram. Ges.*, **40**, 1963, p. 287.
D. A. HOLDRIDGE, *Trans. Brit. Ceram. Soc.*, **64**, 1965, p. 211.
G. JACKSON, *Introduction to Whitewares*, Elsevier Applied Science Publishers, 1969.
P. S. KEELING, *Trans. Brit. Ceram. Soc.*, **60**, 1961, p. 424.
P. RADO, *An Introduction to the Technology of Pottery*, Pergamon Press, 1969.
F. and S. SINGER, *Industrial Ceramics*, Chapman & Hall, 1963.
B. E. WAYE, *Introduction to Technical Ceramics*, Elsevier Applied Science Publishers, 1967.

Index

Absolute viscosity, 133
Accessory minerals, 195–6
Adsorbed substances, orientation of, 120
Adsorption
  acid dyes, of, 119
  basic dyes, of, 117–18
  definition, 115
  gases and liquids, of, 120–2
  non-aqueous solvents, from, 119–20
  physical, 115–22
  positive, 116
Ageing effect, 115
Ageing of clay suspensions, 144–6
Agriculture, 165–6
Alabaster, 227
Albite, 45
Alkali-bearing minerals, 222–5
Alkali metals, 7
Allophane, 35–6
Alumina, 94, 173, 182, 204–8
  chemical properties, 206–8
  occurrence, 205–6
  physical properties, 206
  structure, 204–5
Aluminosilicate minerals, 35–6
Amines, adsorption of, 108
Amphiboles, 28
Anatase, 72
Anauxites, 35
Andalusite, 208
Anhydrite, 227–230
Anion exchange, 109
Anorthite, 45

Antigorite, 31
Atomic absorption, 177–8
Attapulgite, 44
Atterberg plasticity index, 154–7
Atterberg values, 159

Balancing ions, 93
Ball clays, 61–9
  cation exchange, 68
  chemical composition, 64–6
  deflocculation, 68
  dry strength, 67
  extraction of, 63–4
  fired colour, 68
  minerals present in, 181
  modulus of rupture, 67
  occurrence, 61–3
  particle size of, 66–7
  plasticity, 69
  soluble salts, 68
  specific surface area, 67
  typical analysis of, 181
  vitrification, 68–9
  wet-to-dry shrinkage, 67
Basal spacing, 189
Basalt, 49
Bauxite, 205–6
Bayer method, 205
Bentonite, 87
Bingham
  behaviour, 148
  law, 133, 141, 147
  line, 138

Biotite, 44
Bituminous coal, 54
Bloating, 77
Bone, 225–7
Bottom curvature, 137–8
Boulder clays, 80
Bragg's law, 188, 189
Bravais space lattices, 3
Brick clays, 78–87
  composition of, 80–1
  deflocculation, 81
  extraction of, 85
  fired colour, 84–5
  firing shrinkage, 84
  occurrence, 78–80
  particle size distribution, 81
  physical properties, 83
  vitrification, 85
  working moisture content, 81
Bricks, analyses of, 82
Brongniart's formula, 143
Brownian motion, 92, 94
Brucite, 30
Burton U-tube, 100–1

Carbonates, 195
Casting slips, 110, 142–4
Cation exchange, 115
  ball clays, 68
  cause of, 105
  China clays, 60
  reactions, 103–8
    with organic ions, 108
Cation exchange capacity (c.e.c.), 105–7
Cellulose, 52
Celsian, 45
Centrifuge, 179
Ceramic raw materials, range and scope of, 1–2
Chain structures, 28–9
Chemical analysis, 177–85
China clays
  cation exchange, 60
  chemical analysis, 55
  colour, 61
  composition, 58–9
  deflocculation, 60

China clays—*contd.*
  firing shrinkage and vitrification, 61
  location of deposits, 56
  occurrence, 56
  particle size, 59
  plasticity, 59–60
  sources of, 61
  unfired strength, 60
China stone, 222–4
Chlorite-illite, 46
Chlorite-vermiculite, 46
Chlorites, 42–3, 191, 195
Chrome-magnesite, 217
Chrome ore, 215–18
  chemical properties, 217–18
  occurrence and uses, 217
  physical properties, 217
  structure, 216
Chrysotile, 31, 34
Civil engineering, 164
Classical analysis, 177
Clay minerals. *See under* specific types
Clay–water systems
  properties of, 89–126
  rheology of, 127–46
Clays
  classification of, 55
  identification and characterisation of, 177–203
  occurrence of, 55–87
  purification of, 179
Coesite, 21, 23
Colloidal properties, 89
Colloidal solution, 89
Colloidal suspensions, 92–4
Colloids, 89–102
  electrokinetic properties, 92–4
  general properties, 91–2
  lyophilic, 91, 92, 95
  lyophobic, 91, 92
  preparation of, 91
  stability of, 94–102
Compression plastometer, 154
Condensation methods, 91
Cone-and-plate viscometer, 131, 147
Constant rate period, 168
Continuous phase, 90

Co-ordination number, 11–13
Cordierite bodies, 210
Cornish stone, 224–5
Corundum, 204–6
Couette, 130
Counter-ions, 93
Covalent bonds, 8, 9
Cristobalite, 16–20, 22, 173, 175
Critical moisture content, 168, 170–1
Crystal lattices, 7
Crystal symmetry, 4–6
Crystal systems, 5
Crystallinity
  determination, 190
  index, 35
Crystallographic axes, 3, 6
Crystals, 4–6
  bonding in, 6–13
Cubic close packing, 16

Deflocculants, 161
Deflocculation, 110–15
  ball clays, 68
  brick clays, 81
  China clays, 60
  fireclays, 76
Deformation–time curve, 152
Dehydration studies, 121
Devitrification, 21
Dewatering processes, 160
Dickite, 32, 33, 159, 174, 194
Differential thermal analysis, 192–6
  plasters and gypsums, 230
Diffusion, 90, 94
Dilatancy, 134
Dioctahedral structures, 31, 38
Dipole molecule, 9
Dipole moment, 9
Disperse phase, 89
Dispersion methods, 91
Dithionite method, 186
Dolomite, 214–15
Drilling fluids, 163–4
Dry strength, 171–2
Drying, 167–72
Dyes, adsorption of, 117–19

Edge charges, 111–12
Efflorescence, 81
Einstein equation, 139
Elastic flow, 152–3
Electrodialysis, 123
Electrokinetic potential, 96
Electrolytes, 7
Electron clouds, 11
Electron microscopy, 200–1
Electronegativity, 7–9
Electronic structure, 7
Electrons, 8–9
Electrophoresis, 92–3
Electrovalent bond, 6–7, 9
Electro-viscous effect, 139
Emerald, 28
Epigenic agencies, 48
Exchangeable cations, 103–4
Exfoliation, 44
Extrusion, 162–3

Falling rate period, 169
Feldspars, 42, 45, 46, 56, 220–2
Fick's law, 90
Filter-pressing, 160
Fireclays, 69–78
  bloating, 77
  composition of, 72
  critical moisture content, 73
  deflocculation, 76
  extraction of, 72
  firing shrinkage, 76–7
  occurrence, 69–78
  particle size distribution, 73
  plasticity, 78
  properties of, 74
  rational analysis, 72
  refractoriness, 76
  sources of, 71–2
  unfired strength, 73
  vitrification, 77–8
Flame photometry, 177
Flint, 25
Flocculation, 95, 110–15
  degree of, 138
Flow curves, 132, 138, 148, 159
Flow properties, 143

Fluidity, 127, 161
Fluorine atom, 7
Fluorite, 224
Fluxes, 220
Framework structures, 45–6
Freundlich adsorption isotherm, 103
Freundlich isotherm equation, 116
Fuller's earth, 87

Ganister, 24
Geological systems, 50
Geology of clays, 48–88
Gibbs equation, 119
Gibbs isotherm, 116
Gibbsite, 30
Glass, 21, 22, 174, 225, 227
Granite, 49, 57, 224
Group structures, 27–8
Gypsum, 227
  extraction of, 230–1

Halloysite, 33, 34, 123, 160, 174, 194
Hashimoto and Jackson method, 186
Heat effects, 167–76
Hectorite, 38
Helmholtz-Smoluchowski equation, 102
Hexagonal close packing, 16
Hofmeister Series, 103
Hooke's law, 150, 152
Hornblende, 29
Humic acids, 53, 54
Hydrated form, 33
Hydrated oxides, 196
Hydration effect on ionic radius, 98
Hydrogen atoms, 8, 9
  bonding, 10, 33
Hydrogen clays, 123–6
Hydrosphere, 95
Hydrous mica, 41
Hydroxonium ions, 42
Hypogenic action, 49
Hysteresis effect, 136

Igneous rocks, 48
  composition of, 49, 51

IL/MA test, 187–8
Illite-montmorillonite, 46
Illites, 35, 41–2, 175
Imogolite, 35–6
Impurity effects, 174
Infra-red absorption spectrophotometry, 35
Infra-red absorption spectroscopy, 197–9
Intercalation compounds, 122–3
Interstratified structure, 46
Ion exchange effect on viscosity, 141–4
Ionic formulae, 183–5
Iron oxide removal, 186
Island structures, 27
Isoelectric point (IEP), 94
Isomorphous substitution, 28, 29, 34, 36, 39, 41, 42–4, 46

Kaolinite, 32, 33, 56, 64, 102, 156, 159, 172, 173
  characteristics of, 193
  differential thermal analysis, 192
  disordered, 33–5, 42, 68, 194
  separation method, 186
  well-crystallised, 35, 159
Kaolins, 30–6, 55
  characteristics of, 189–91
  deflocculation and flocculation, 110
  plasticity, 159
  stacking in, 32–3
Keatite, 21, 23
Kyanite, 208

Langmuir isotherm equation, 103, 116
Lattice structure, 11
Lattices, 2–3
Leather-hard moisture content, 169
Lepidolite, 41, 222
Lignin, 53, 54
Lignite, 65
Limit of irreversible adsorption (L.I.A.), 118
Linear shrinkage, 171
Liquidity index, 164

Index

Lithium ion, 7
Lyosphere, 95
Lyotropic series, 103

Magma, 48
Magnesite, 211–14
Magnesium oxide, 213
Margarite, 40
Memory of clays, 153
Meta-form, 33
Metakaolin, 173
Metamorphic rocks, 48, 54
Methylene blue, 117–18
Micas, 39–41, 191, 195
   dehydration, 175
Miller Indices, 6
Mineraliser, 18
Mixed-layer structures, 46
Moisture content, 167–9
Montmorillonites, 34–9, 56, 113–15, 148, 154–6, 159, 183–4, 191, 194
   adsorption, 120–1
   chemical analysis of, 184
   differential thermal analysis, 193
   intercalation of, 123
   thermal decomposition, 175
Mössbauer spectroscopy, 201–2
Mullite, 173–5, 208, 209
Muscovite, 39–42
   dehydration of, 175

Nacrite, 32, 33, 159, 174, 194
Negative charge, 37
Nepheline, 222
Nepheline syenite, 224
Nernst potential, 95
Newtonian fluid, 163
Newtonian liquid, 129
Newton's law of viscosity, 102, 129, 130, 132
Non-aqueous solvents, 99, 119
Non-Newtonian flow, 132–46, 163
Nontronite, 37
Normal spinel, 216

Olivines, 27
Optical microscopy, 199–200
Organic matter, 196
   removal of, 187
Orientation of clay particles, 135, 171
Orthoclase, 45
Overburden, 63, 85
Oxygen atoms, 8, 17

Paint industry, 164–5
Palygorskite group, 43–4
Paper-making industry, 165
Paragonite, 40
Pauling's rules, 11
Pegmatites, 224
Periclase, 213
Petroleum industry, 163–4
Pfefferkorn index, 157
pH
   effects, 24, 94, 111
   titration curves, 123–6
Phlogopite, 39, 44
Physical adsorption, 115
Pipe-clays, 87
Plaster of Paris, 227–31
Plastic flow, 133–5, 147–9
Plasticity, 147–66
   definition of, 147
   of clay minerals, 159–60
Plasticity index, 154
Plasticity theory, 157–60
Point groups, 4
Polar liquids, 95
Polar molecule, 9
Polarity of bonds, 9–10
Polymorphism, 41
Positive charge, 37
Potash-bearing minerals, 220–5
Potash mica, 39
Pressure–flow curves, 148, 149
Proximate analysis, 180
Pseudo-plasticity, 134
Pyrites, 196
Pyrophyllite, 36, 39, 175
Pyroxenes, 28

Quartz, 15, 18–19, 22, 195
 determination of, 186
Quartzites, 25

Rational analyses, 180, 182–3
Refractory raw materials, 204–19
Residual clays, 55–61
Residual deposits, 49
Resonance, 10
Rheology
 definition, 127
 of clay–water systems, 127–46
Rheopexy, 138–9
Rock types, 48
Rotating-cylinder viscometer, 130
Rutile, 21, 218

Salts, 7
Saponite, 38
Scaffold structure, 122
Sea-water magnesite, 211–13
Sedimentary clays, 55
Sedimentary rocks, 48, 49
 composition of, 49–55
Sedimentation volume, 113
Selenite, 227
Separation methods, 185–8
Sepiolite, 44
Sericite, 41
Serpentine, 31, 215
Shale, 52, 55
Shear-hardening, 151, 152
Shear rate, 129
Shear stress–shear rate curves, 139
Shear-thickening, 139
Sheet structures, 29–30
Shrinkage factors, 170–1
Silcrete, 25
Silica, 14–26, 173
 action of fused alkalis on, 24
 amorphous, 20
 chemical properties, 23–4
 conversions, 17–18
 crystalline form of, 15, 17, 24
 decomposition for analysis, 24
 forms of, 20–2

Silica—contd.
 gravimetric determination of, 23
 high pressure forms, 21, 22
 inversion of forms of, 18–20
 occurrence of, 24–5
 physical properties of, 22–3
 polymorphic, 14
 thermal expansion of, 23
 vitreous, 21
Silica gel, 20, 23
Silica W, 21–2
Silicates, 208–9
 soluble, 24
 structures, 27–30
Silicon, 14
Silicon-oxygen tetrahedra, 14, 16
Sillimanite, 208
Sintering, 206
Slate, 52, 55
Slip-casting, 142–3, 160–2
Smectite, 39, 113–15
Soda-bearing minerals, 220–5
Sodium silicate, 24
Soil
 constituents, 165
 mineralogy, 164
Soluble salts, 68, 81
Solvation, 95
Space groups, 5
Spectrophotometry, 177–8
Spinel, 215
Spodumene, 222
Stacking in kaolins, 32–3
Steatite, 209–10
Stern layer, 95
Stern potential, 95, 96
Stern theory, 95, 99
Stishovite, 21, 23
Stokes' law, 90, 179
Stoneware clays, 87
Strain-hardening, 151
Streaming potential, 100
Streamline flow, 128
Stress–strain
 curve, 150, 151, 153
 measurements, 149–54
Structure, fundamental principles of, 1–13

Swelling of clays, 87, 114
Symmetry, 4–6

Talc, 39, 210
Tetrahedral layer, 31
Tetrahedron, 14
Thermal decomposition, 172–5
  montmorillonites, 175
Thermogravimetric analysis (TGA), 196–7
Thixotropic loop, 137
Thixotropic structure, 131
Thixotropic suspension, 131
Thixotropy, 135–8, 161, 163
  cause of, 136–7
  definition, 135
  degree of, 137
  negative, 145
Titanium dioxide, see rutile
Torsion viscometer, 131
Tridymite, 16–17, 20, 22
Trioctahedral structure, 38
Turbulent flow, 128
Tyndall effect, 92

Ulmins, 53
Ultrasonics, 144–5
Underclays, 69
Unit cells, 3–4
Unit lengths, 6

Valency forces, 11
Van der Waals bonds, 11, 46

Van der Waals forces, 115
Velocity, distribution of, 128
Velocity gradient, 129
Vermiculites, 44
Virtual yield value, 137–8
Viscosity,
  absolute, 140
  definition of, 127
  effect of ion exchange, 141–4
  effect of solid concentration on, 139–40
  measurement, 129–32
Viscosity coefficient, 129
Vitrification
  ball clays, 68–9
  brick clays, 85
  fireclays, 77–8

Water-glass, 24
Wave mechanical theory, 10
Wollastonite, 222

X-ray diffraction, 2–3, 33–6, 46, 186, 188–91

Yield stress, 133
Yield value, 133, 138, 152

Zeta-potential, 97, 99
  determination of, 99–102
Zircon, 219
Zirconia, 219